"What's wrong with my plant?"

A
Guide to
the Care and
Cure of Ailing
Houseplants

Chuck Crandall

Chronicle
Books

Second printing, February 1977

Library of Congress Cataloging in Publication Data

Crandall, Chuck.
   "What's wrong with my plant?"

   Includes index.
     1.   House plants—Diseases and pests.   2.   Pest
control.   3.   House plants.   I.   Title.
SB608.H84C7   635.9'65   76-19086
ISBN 0-87701-080-3

Chronicle Books
870 Market Street
San Francisco, Ca. 94102

contents:

Chapter 1
good beginnings

If you own more than one plant you need this book. It is not simply a plant-doctoring guide, it is a plant-care manual. Your goal should be to keep your plants healthy and thriving so they never need "hospitalization." Let's say that what all plants need is preventive medicine. A plant that has never been invaded by parasites or disease is always a stronger, more attractive specimen. Parasite or disease invasions inevitably weaken a plant and sap some of its vigor. And usually some of the foliage is destroyed before control is achieved. Once a plant has been weakened by parasitic assault or disease it seldom regains its original strength. It may be subject to relapses or be more susceptible to new plagues than a plant which has never been invaded.

So you begin to see why the emphasis should be on prevention, not treatment.

Check New Plants Carefully Before You Buy

More often than seems fair, plants with problems are unwittingly brought home from the neighborhood supermarket or nursery. When foliar damage or deterioration is discovered, you aren't sure whether the plant was sick when you bought it or "got something" after you took possession. Sometimes it's tough to determine which is which.

Obviously, you should buy only vigorous, healthy plants. Now, it is all well and good for someone to tell you to make sure the plants you buy are healthy and pest-free, as if you intended to seek out diseased or infested specimens. But there just is no 100% sure-thing when you're plant-shopping. Buying plants is a gamble.

A well-intended but worthless piece of advice frequently given is to insist on a guarantee from the retailer when you buy a plant. It is implied that if the retailer is reputable and honest he'll have no objection to standing behind what he sells. Few

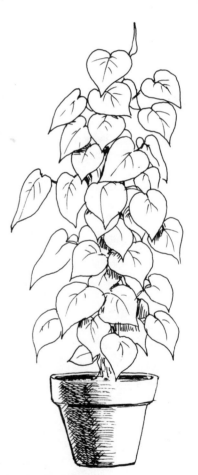

How To Pick A Plant *Look for thick, lush foliage on a stout stem or trunk, with no evidence of leaf damage or deterioration. Foliage should be uniform in color and size.*

plant-sellers *do* go this far, and those who do have more compassion than business sense. Plants are living things and are easily damaged or killed by carelessness, neglect and bad horticultural treatment. Once the plant leaves the shop, the owner has no way of knowing what the purchaser has done to it. It may have been left in the back of a closed station wagon to cook in the sun while the new owner ran errands all over town Or it may have been drowned with water on a daily basis for two weeks until root rot set in. Or it may have been given a year's supply of fertilizer in one day.

It simply isn't fair to expect the plant-seller to guarantee his stock. He'd soon go broke if he did, and for this reason few do. That's why it's important for you to examine a plant carefully to know what you're getting before you buy.

There are obvious trouble signs to look for when you're plant shopping; these will be covered a little further on. But most of the pests which attack houseplants are microscopic—you can't see them even if you look for them. Many plant wholesalers (and some retailers) spray their plants with a pesticide if they know their stock is infested, but pest eggs are not always wiped out along with the adults, especially those laid inside leaves. So you buy the plant and take it home to a hot, dry environment that is an ideal "incubator" for hatching these eggs into a colony of destructive pests. Sounds really unfair, doesn't it—having to pay for the privilege(?) of fishing someone else's chestnuts out of the fire, but— fortunately—this doesn't happen too often. Most growers and retailers go to great lengths to ensure that their stock is pest- and disease-free.

Except for hitch-hiking insect eggs, most other signs of sick, infested plants are easily recognized in most species. In most cases, foliage is the key sign of plant health and vigor. It should be a rich, lush green, cool to the touch and standing out slightly from the stem. There should be no uncharacteristic gaps between leaves or leaf tiers on the stem. Widely spaced leaves or tiers on a weak, spindly stem or trunk usually indicate the plant was poorly grown or given excessive amounts of fertilizer to bring it along fast so it could be delivered to the retailer sooner. The result is a trade-off—height for fullness. A short, full plant will gain height in time and grow into a beautiful specimen; a tall, sparse-foliage plant will never fill in the naked areas and will always be thin and gangly—unless you cut it back and let it start again. But why should you have to endure this?

Missing leaves and evidence of trimmed foliage edges are two more signs of a damaged plant. Some retailers trim off brown leaf tips and margins or dead and dying leaves to make the plant more attractive and desirable to a prospective buyer. I doubt that this is a deliberate attempt to hoodwink anyone, but it isn't a very commendable practice. Such shorn and pruned specimens should be offered at a substantial discount but rarely are, so check the foliage of a plant you're thinking of buying for indications of a recent "haircut."

The browned-off margins and tips and evidence that leaves have been removed probably don't mean the plant was infested or diseased. Several things could be responsible—for example, low humidity, fertilizer burn, sunburn, dehydration and rough handling. But you should expect to get a healthy and attractive specimen when you plunk down your money.

All pests which attack foliage and blooms are visible under close scrutiny, or the damage they cause is evident and identifiable. Refer to the section on pests and how they affect a plant for specifics, but—as a general rule—steer clear of plants with brown, yellow or red flecks or spots on the foliage, and those whose leaves have soft, white, mushy areas which indicate the presence of a fungus disease. Flowering plants with deformed buds and blooms or with buds that drop off at the slightest touch should also be passed by.

A hole chewed through one or two leaves or an occasional yellowing leaf, especially on the lower part of the plant, are no cause for alarm. Insects which eat through a leaf usually nip and run from plant to plant. Chances are good they're not still chomping away on the plant, but if they are they're large enough to see and pick off. A lower leaf that is yellowing or browning-off is a natural phenomenon in plant physiology. In a healthy plant, the lower leaves are always the first to die as a plant adds new leaves at the growing tips of higher branches and transfers to them the energy and food it would have used to nurture the lower leaves. In some specimens, the natural tendency of a plant to sacrifice lower leaves for new growth can be frustrated by pinching out or cutting back the growing tips. Energy is then returned to sustaining older foliage which enjoys a sudden burst of growth, until new foliage develops at the growing tips again.

Another important factor to consider before you buy an expensive plant is its environmental requirements and whether you can duplicate these in your home. Don't assume that because the plant you're considering is a "house" plant that it will continue to thrive wherever you decide to put it indoors. For example, some plants require filtered or direct sun to survive. *Ficus benjamina*, cacti and other succulents and most flowering houseplants are a few with the filtered/direct sun requirement. If your home or apartment gets very little sun or very poor light, you wouldn't want to invest $75 to $125 in a

Be a fussy, observant plant shopper.

Ficus benjamina tree only to see it deteriorate over the next few months.

So if you don't know what kind of environment a particular specimen requires, talk to the retailer. It is far better to display a lack of knowledge than to throw away your money and kill a beautiful plant in the process. An alternative would be to get the proper botanical name, go to the library or a large bookstore and look up the plant in *Exotica* or some other horticultural reference work.

After You Buy a Plant

Concern for your new plant's welfare should begin the moment you walk out of the shop with it. The simple process of moving it from the store and driving home with it is fraught with potential dangers. If it's a very hot day you should have the salesman mist the foliage before you start out, and if you have a long way to go and you don't have auto air conditioning, open a window so the plant won't keel over from heat prostration. Don't, however, let foliage hang out the window where it will be whipped and shredded by the wind. If you stop for more than a few minutes, crack all windows at least an inch—more, if it's safe to do so without risking a rip-off by plant-nappers.

If the air is chilly or wintry, take the opposite precautions. Keep the plant warm and protect it from cold drafts. It can cope much better with a stuffy, overheated car for the trip home than with being exposed to chill and icy drafts.

When you get the plant home, don't leave it outside in the sun for even a few minutes. Its foliage has never been struck by the direct rays of the sun and can be severely burned. If the weather's cold, get it inside quickly to avoid shock.

Once you've safely arrived with your new plant there are two steps you should take before relaxing and enjoying the adoptee. First, pour about a quart of lukewarm water into the pot and wait for a few minutes, then pour in another quart. The first quart dissolves any concentration of mineral salts (fertilizer residue) and the second helps flush the salts out of the pot to prevent a build-up and leaching through the pot, which is evidenced by white stains on clay containers. If the plant is in a plastic pot, this safeguard is even more important. Growers use considerably more fertilizer than you should to bring the plant along. Sometimes a plant will not use up the extra doses and the excess may not be flushed out with normal irrigations. The problem is compounded if you unknowingly add even more fertilizer later on.

The second step is to isolate and watch the plant for a week to ten days to make certain it's healthy and pest-free before moving it in among your other specimens. Let the soil dry out during the incubation period. This will encourage the feeder roots to go looking for moisture and this root activity will help invigorate the plant and overcome any trauma it may have suffered from the sudden change in environment. If the weather is hot and dry, mist the foliage twice a day to keep the humidity high.

If you're a worrier, or you have some reason to suspect that a plant you've just acquired is harboring destructive critters, you can give the newcomer a "delousing" insurance spray of suds from an Ivory soap bar, followed by a clear water rinse. (Cover the topsoil with aluminum foil to keep the soap solution out of the pot.) Ninety-nine times out of a hundred, this is sufficient. Insecticides shouldn't be used indiscriminately, but only in response to a real need. After the incubation period, if the plant seems to be thriving and free of predators, introduce it to your collection.

Problems with new arrivals are covered under the heading, "Shock," in the chapter on plant diseases, causes and cures.

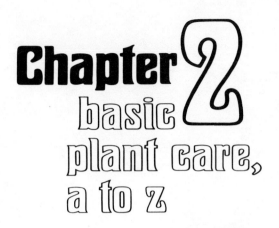

Chapter 2
basic plant care, a to z

Scene:
A midtown apartment, Anywhere, USA. Two people are standing over a plant that appears on the verge of collapse.

Plant-owner's Friend:
Gee, your whatchamacallit plant looks a little sick.

Plant-owner (Concerned):
Yeah, really! I'm afraid it's going to die.

Friend:
Well, are you watering it?

Plant-owner:
Sure! Every day.

Friend:
Well, I guess it isn't thirsty, then. What about plant food? Maybe it's hungry.

Plant-owner:
Gee, I don't think so. I've been feeding it regularly every three days.

Friend:
Whatcha been feeding it?

Plant-owner:
Everything. Vinegar, egg shells—that kinda stuff. You know, I'm heavy into organics.

Friend:
Egg shells?

Plant-owner:
Sure. They're a great source of calcium—or something. I forget.

Friend:
Well, why don'tcha put it outside for a while? Maybe the sun'll cure it.

Plant-owner:
Yeah, I guess I should try that. I've done just about everything else to it.

Fade Out

Correct Planting Depth *Plants should be potted up so that there is about an inch between the topsoil and pot rim. Plant on top is too high, plant in center is too low, plant below is proper depth.*

The preceding scenario may seem a bit exaggerated to illustrate the point that we are often our plants' worst enemies, but I've met plant-lovers(?) who have done all of these things to their plants. Some, I'm sure, are still killing off plants on a regular basis, even after my remonstrances. There is nothing so impregnable as a closed mind.

A lot of people believe there is something mysterious or esoteric about growing healthy plants and trees indoors. Not true. It requires no particular skill, knowledge or special gift. Not particularly bright kids can do it, and I suspect even a chimp could be trained to care for a philodendron. Too many people insist on making a ritual of it, employing everything from astrology to Zen in their quest for the secret of the Green Thumb.

The secret of the green thumb, if it exists, is simply *Give your plants what they need to thrive—nothing more, nothing less.* Sounds easy enough, but what does it mean? Some plants require more water than others. Some like the sun streaming through a window, some are sensitive to direct solar rays. Some like dark corners, others wither and die in a gloomy location.

You'll find some fundamental guidelines on what individual species require later on in this book, but you'll have to do some research on your own to determine what environment and specific care a particular kind of plant needs. The best place to do this research is at the shop or nursery where you buy your plants or in *Exotica*, the definitive horticultural reference book on tropical and subtropical plants. Try your library or the largest bookstore in town.

What you will learn in *this* book are the basics—what all plants need, in general, to survive and thrive indoors, how they live and grow and what's wrong with them when they don't.

What are some of the things plants require? Light, water and air are three of the major requisites. Most also need moderately high humidity to help them cope with the dry, hot air found in most of our homes, and some nutritional help.

Light

Let's take light first. Anyone who once had a semester of biology knows that plants need light. Light striking the surfaces of a plant's leaves helps the plant manufacture food for growth. The process, you'll recall, is called photosynthesis. Usually, the deeper the green of a plant's foliage, the more light it requires. But all foliage plants need some natural or artificial light to live. No foliage plant can survive very long in darkness.

Inadequate illumination can cause a plant that may be getting all the other things it needs to "go leggy" (widely-spaced, stunted leaves on the stem), to "pale out" (lose its rich, deep-green color and appear washed out), or to sit there month after month simply holding its own and adding little noticeable new growth.

Even plants sitting directly in front of windows are getting light from only one direction. As a result they tend to lean toward the light and branches facing the window produce more and healthier foliage. For this reason, you should rotate plants that are situated near a natural light source half a turn every four to seven days. This encourages fullness throughout the plant and also keeps it growing straight.

If you live in an apartment or house that gets poor natural light and no direct sun at all, you either will have to settle for small plants and those with minimal light requirements, or will have to invest heavily in artificial light setups, such as Gro-Lux and Gro-Lux Wide Spectrum (Sylvania), Plant-Gro (Westinghouse), or Plant-Lite (General Electric), all of which provide sufficient illumination for small to medium-sized plants. At present, artificial light equipment designed for growing plants is still virtually in the developmental stage and is accordingly expensive, but the situation is improving yearly.

Those who are desperate for really showy foliage plants and trees, and are indifferent to the appearance of commercial fluorescent light fixtures suspended from their ceilings, can go this route. Fluorescent light almost approximates daylight illumination (without the glare) and nearly all plants, especially large-foliage specimens, thrive in it. Another plus is that because it is cool light it poses no danger of burning or dehydrating plants.

Which brings us to traditional incandescent bulbs, the type used for normal household illumination. Incandescent light can be used, at a distance, to supplement the natural light, but it is too limited in candlepower and area of coverage to be very useful on all but small windowsill or table-top plants. Still, it is better than nothing. Care must be taken to make sure the bulb is far enough away from foliage (at least eight inches) to prevent burns or drying out.

Some Plants
and their Light Requirements

Low Light, North Windows

Aechmea fasciata
Anthurium
Araucaria heterphylla
Asparagus ferns
Aspidistra
Cissus (nearly all)
Crassula argentea
Dracaena
Fatshedra
Ficus elastica
Maranta
Philodendron
Sansevieria
Spathiphyllum

Bright-Diffuse Light, Near East/ South Windows

Aralia
Asparagus ferns
Begonia
Camellia
Cissus
Clivia
Codiaeum
Coffea
Cordyline
Crassula
Cyclamen
Dieffenbachia
Dracaena
Fatsia
Fuchsia
Gardenia
Gloxinia
Howea
Philodendron
Schefflera

Some Direct Sun, Eastern, Southeastern & Western Exposures

Abutilon
Agave
Aloe
Amaryllis
Ananas comosum
Azalea
Bambusa
Beaucarnea
Bougainvillea
Cacti
Carissa
Chrysanthemum
Citrus
Coleus
Crassula
Euphorbia
Gardenia
Kalanchoe

You will notice that some plants are on two or all three lists. Most of those on the low-light requirement list will do much better in brighter light but will survive with dimmer illumination.

Plants situated in sunny windows may need some protection from both the heat build-up (sometimes as high as 120°F on midsummer days) and the direct rays of the sun to prevent scorching of foliage and dehydration. Cross ventilation or air conditioning will solve the first problem and a thin curtain or shade between the plant and the pane is usually adequate in solving the second. Cacti and most other succulents seem to cope just fine with undiffused sun.

Water

Water carries a plant's food supply and constitutes 90% of a plant's composition. An adequate supply of water keeps the leaves cool and turgid (at their proper thickness) and provides sustenance for the plant. Too little water causes a plant to deteriorate from dehydration—but too much water is the kiss of death.

Overwatering is responsible for the demise of more plants in homes all over the country than all other problems combined. There is a persistent popular belief that plants need water every day if they are to thrive. At the conclusion of a lecture on plant care recently, during which I explained proper watering procedures, a waspish little woman confronted me as I was leaving the dais. "Young man," she began, wagging a minatory finger in my face, "obviously you don't really understand or love plants. I water mine every day and, if they're not thirsty they don't drink it. But, it's there for them every day, in case they want it." There was nothing I could say in the face of such overwhelming logic. I do hope, though, that this well-intended Lucrezia Borgia of plantdom will soon shift her interest to ill-tempered lapdogs.

The kind of water you give your plants is important only if your local water supply is loaded with heavy concentrations of chlorine. If so, and your collection of plants is small, you should switch to bottled drinking water. A gallon sells for about 49¢ at the supermarket. Obviously, if you are growing several large plants or trees, the cost of bottled water would be prohibitive. In this case, draw tapwater the night before you plan to irrigate and let it stand in the containers. This will allow some of the chlorine to evaporate.

If you live in suburban or rural areas, away from the polluted air of the big city and industrial complexes, and it is practical to do so, you should collect and store rainwater to use in irrigating your plants. Rainwater, of course, contains none of the chemical additives which are potentially harmful to plants. It can be stored for months in tightly capped plastic bottles or gallon jugs.

Rainwater from urban areas is not recommended since it probably contains a high concentration of pollutants from auto emissions and industrial waste, which are extremely harmful to plants. There is no practical way to filter these impurities from rainwater to make it safe.

Root shock can occur in plants if the temperature around the roots is changed suddenly. For this reason you should never pour very cold or hot water into a plant's pot. Wait until it is room temperature or even a little on the lukewarm side. Roots take up warm water faster than cool.

Watering Plants: When and How.

As a general rule of thumb, most of the plants we grow indoors need water only about every seven to ten days. The exceptions are cacti and other succulents which require soil moisture sometimes as seldom as once a month, and a few of the water-loving plants, such as Spathiphyllum, asparagus ferns, Chlorophytum comosum (ribbon plant), to name three, which may need water as frequently as every two or three days.

A number of factors determine when a plant requires moisture. Soil in plastic pots stays moist two to three times longer than soil in clay pots. Plants grown in an overheated environment or in a sunny or partially sunny window use up more water faster than those grown in cool environments or areas that get only bright to medium light. On hot, dry days plants transpire (exude water vapor) more than on cool, overcast or humid days, so their lost moisture must be replaced more often. With all of these variables at work, there is no hard-and-fast rule about when to water. This is why watering schedules, in which you irrigate all your plants on a given day each week, are useless.

Ninety percent of all houseplants are tropical or sub-tropical species. Fortunately for most of us who like to play with our watering cans, tropical specimens can handle a lot more water than plants from other regions. Most require moisture when the soil is dry two to three inches down, in larger containers. Plants in small clay pots (4″ to 6″) should be watered when the soil is arid to a depth of a quarter inch and small plastic pots when it's dry an inch down.

Feel the soil with your finger before you irrigate any plant. This is your most reliable method of gauging the moisture content. Don't merely *look* at the topsoil. Because the air in the

A finger in the soil is the best way to determine soil moisture/aridity.

interior of your home is probably hot and dry, the top layer of soil is probably also dry, but further down the soil may be adequately moist.

Never keep the soil continuously wet. Swampy soil is an open invitation to fungus disease and root rot. Boggy soil interferes with a plant's ability to function. When the soil begins to dry out, the feeder roots are stimulated to push through the soil to look for moisture. The longer the dry spell, the more frantically they search. Drying soil shrinks slightly, allowing vitally needed oxygen to enter. All of this is what keeps feeder roots and, ultimately, the entire plant healthy and thriving. Obviously, if the soil isn't given a break from the deluge of moisture, the feeder roots go nowhere. They sit with their feet in the swamp day after day until they begin to rot.

This is why it is important to give a plant water only when it needs it, and to make sure the excess is passing out the drainage hole where it can be dumped from the saucer. Drainage holes should be kept clear so that the water evacuation process is achieved each time you irrigate. Crocking over the drainage hole helps keep obstructions from blocking the escape hatch.

Plants in containers without drainage holes should be laid on their sides and tilted in the back slightly for ten to fifteen minutes so that excess water can drain out.

All of this talk about wet soil may seem somewhat overdone, but just remember that a plant which has been underwatered can usually be saved; a plant which has been

Ample crocking over drainage hole(s) keeps roots from clogging holes.

overwatered and has developed root rot seldom can be salvaged.

For some obscure reason, the simple process of irrigating plants creates a great deal of controversy among plant enthusiasts. Wick-watering is in vogue again, and many people are also dumping trays of ice cubes on the topsoil of their plants and letting them melt so that the plants get a gentle trickle of water. Well, wick-watering provides enough moisture for dish-garden plants but is woefully inadequate for plants in large containers. The ice cube method poses a serious threat to the well-being of a plant. Most plants have roots just under the surface of the soil. If these, or the stem, become frozen, shock could easily be the consequence. The theory in both methods seems to be that you have to sneak up on the plant with moisture: if you walk right up to it with watering can in hand and say, "Alright, fella, I'm going to water you now," it will keel over from a cardiac arrest.

There is absolutely nothing wrong with overhead watering—in fact, it's the best way to irrigate tropical and subtropical specimens. The soil—all of it—needs to be thoroughly soaked on a regular basis, then allowed to dry out slightly between irrigations. Not only does this keep the soil oxygenized and fertile, but the countless rootlets growing throughout the soil ball benefit immensely from moisture. The myth that overhead watering is harmful probably got started in the Victorian era when nearly all plant fanciers grew Boston ferns. Water poured directly in the center of Boston ferns often causes the fronds at the core to die. However, this can be avoided by watering these plants around the edges, just inside the rim.

Since a plant's root system extends throughout the soil ball, pots should be completely submerged in buckets or tubs of water every two weeks from spring to fall. This has several beneficial effects. First, it invigorates and revitalizes the root system; second, it collapses the tunnels and canals that have been carved by irrigation water in the soil, so moisture will spread more evenly in the future and not rush down through

Bucket-soaking thoroughly moistens soil and roots.

these tunnels; and third, it helps break up caked and compacted soil which is brought about by consistent overhead watering.

You'll know the soil ball is completely soaked when air bubbles no longer rise to the surface. These result from tunnels and pockets in the soil ball that fill in after the operation. Allow pots to drain well after soaking and return them to their saucers.

Air.

Oxygen is essential to nearly all living things. Plants need oxygen in small amounts to carry on the process of photosynthesis. Obviously, then, good ventilation is necessary to the health of plants. The temperature of the air is an important factor, as well. Tropical and subtropical plants, which, as we've seen, constitute most of the houseplants we grow, can cope with heat better than with cold, but problems occur for them in closed environments where the air temperature rises above 75°F. Overheating "cooks" the oxygen out of the air and dries out the foliage of plants, causing them to transpire excessively.

If, for reasons of security, heavy atmospheric smog, etc., you can't leave windows open on hot summer days, especially when you're gone on vacation, or if you don't have the luxury of air conditioning, use an oscillating fan near your collection to keep the air moving and help cool it.

In winter the problem of providing fresh air for our plants is almost insurmountable. Our own comfort comes first and we are not willing to shiver all day by cracking a window so our plants can get a breath of fresh air. We keep the thermostat high, which reduces the oxygen content of the air. There is nothing much that can be done to improve the situation in the dead of winter. Wiping down the foliage with a cool moist cloth helps some. The best solution, if you live in an area where winters are long and severe, is to limit your collection to tough, durable, thick-leaved varieties, such as rubber trees and other ficus specimens.

Cold blasts of air are just as debilitating to plants as is hot, dry, stale air. A plant near a doorway in winter will almost always deteriorate and may even go into shock, dropping leaves or simply keeling over. Plants near windows in winter react the same way unless shades or drapes are drawn against the cold air. Some may even freeze overnight if they were recently watered.

Humidity.

Humidity is, by simple definition, water vapor suspended in air. Tropical plants in the rainforests thrive in a humidity level of 70% to 80%. Houseplants would love this muggy level, too, but you would find it unbearable. Outdoors, there is always a measurable humidity level. On hot, dry days it is of course very low; when it is cool and misty, or warm and sultry, humidity is very high. But indoors, especially on the furnace-heated days of mid-winter, humidity is almost nil.

Moisture in the air around a plant helps keep its foliage cool and fresh and invigorated. Some of the moisture that settles on leaves is absorbed by them through their *stomata,* or pores—provided these are not blocked by dust or layers of grime from smoking and cooking.

Heat is the biggest enemy of humidity. When furnaces are turned on in winter, the humidity level plummets to 10% or lower. While cacti and most succulents find this level satisfactory, most foliage plants prefer a level of 70% but will settle for 30% or 40%. If your plants are suffering from low humidity, you'll notice the deterioration first at the tender, sensitive leaf tips. They turn brown as the cells die, and eventually the edges of the entire leaf begin to brown off. Leaves may also curl slightly and feel crinkly and dry.

If you have a large collection of foliage plants, or are planning to, you should invest in a hygrometer which will enable you to check on the humidity level indoors. Most large nurseries, plant shops and hardware stores stock them and they sell for about $5. This helpful little instrument takes the guesswork out of determining when you need to raise the humidity level around your plants.

Raising Humidity Levels Indoors.

Ideally, every indoor gardener should have a humidifier that is built into the heating/ventilation system. This provides for the automatic injection of moisture into the air and eliminates most of the fussing over plants with misting atomizers. Few of us are so fortunate and must resort to more primitive methods of humidifying the air.

For a quick, temporary treatment, the hand atomizers are hard to beat. The nozzle should be adjusted to produce a fine mist. Coarse drops of water simply run off foliage and create a bigger, messier clean up later on, or—worse—don't evaporate by nightfall, creating ideal conditions for the development of fungus disease.

If the plants are portable, the best place to do the misting is in the shower or tub where run-off and overspray pose no threat of water damage to floors, furniture, books, etc. Here you can also use the shower head instead of the mister. If plants are too heavy to hoist, or you have too many, spread

Hand-misting replaces moisture removed from air by interior heating systems.

newspapers around them to catch any wayward mist. Cover the foliage completely with the fine mist and let it evaporate.

There are a few important precautions to take when you mist your plants. First of all, don't mist hairy-leaved specimens, such as African violets, or cacti. This is an open invitation to mildew and fungus infection. Second don't mist on dull, overcast or warm, muggy days. There is already sufficient humidity in the air. Adding more will only create favorable conditions for fungi growth since the moisture you add probably won't evaporate. For this same reason, you shouldn't mist later than noon or one o'clock, especially on cool days. Finally, if weather permits, open a window or two around freshly misted plants or use a small fan to keep the air circulating for half an hour. Moist, stagnant air is a welcome mat for mildews and fungi.

Humidity can be raised around plants for longer periods by creating a microclimate of moist air. This is accomplished by placing oversized saucers filled with pebbles under pots and pouring a little water into the saucer. The pebbles raise the pot enough so that the water doesn't reach the drainage hole. For days, vapor from the saucer rises under the plant and keeps the foliage refreshed. Bowls, pans or other containers of water can be used to accomplish the same result. Simply slide them in among your collection and from time to time fill them with warm water.

Fertilizers and Plant Nutrition.

During spring and summer, the heavy growth period of most plants, a little fertilizer is beneficial. The operative word is

Create a high-humidity microclimate by pouring a little water over pebbles in a plant's saucer. Rising moist air bathes the leaves continuously, recreating greenhouse atmosphere.

"little." Plant-lovers are a doting, overindulgent breed. There's nothing wrong with this: plants need and appreciate attention. But problems begin when we decide our plants might possibly do better with an added dose of fertilizer. Or, worse, when this thinking is carried a step further: If plants grow well in spring and summer, couldn't they be induced to continue producing new foliage and blooms in fall and winter with regular "booster shots" of plant food?

There are two potentially fatal flaws in this reasoning. First, plants neither need or want extra fertilizer. Even following the dilution recommendations on the label of a container of plant food, we often overdose our plants by dumping these nitrogen-rich concentrates at one fell swoop into their tiny, confining containers. Instead of getting a gentle "buzz" of nutrition over a period of time as they do when growing naturally in the ground, they get a blast all at once. Sometimes when we overdo it the price we pay is plant collapse from root shock or "burn," which means the tiny, delicate feeder roots which take up water and nutrition are singed by corrosive nitrogen particles. Since these vital roots have been damaged and can no longer collect water to sustain the plant, it may die before new feeder roots can be produced.

The second flaw is that nearly all plants go into dormancy for a period between fall and winter. Days are shorter, light

diminishes in intensity and the air is cool. All of these changes induce plants to slow down and rest. They don't require as much water and no fertilizer at all, since they are virtually inactive. The unused food will just sit in the pot building up to dangerous levels and root burn can be the final result. So withhold all nourishment from about September through April.

The type of plant food you use is important, too. It should be formulated for the kinds of specimens you're growing. Acid-loving flowering and foliage plants and all citrus need a high-nitrogen, or acid, fertilizer. Most tropicals prefer a food more or less evenly balanced among nitrogen, phosphorus and potassium. Whatever brand you use, it should be a *complete* fertilizer. To be a complete plant food, a product must have these three major elements, nitrogen, phosphorus and potassium. You can determine whether a plant food is complete simply by looking at the label. All major brands have three numbers printed on them, all in a row: for example, "5-10-5." This means that the food is five percent nitrogen, ten percent phosphorus and five percent potassium or potash. The order never changes. The first number always represents nitrogen, the second, phosphorus, and the third, potassium. A plant food with numbers that read, "0-10-0" is obviously not a complete fertilizer but is 10% phosphorus and 90% inert or inactive ingredients.

Three fertilizers which have proven reliable for the author over the years are Stern's Miracle-Gro (15-30-15), Stern's Miracid (30-10-10) and Germain's Fish Emulsion (5-1-1). The first is a good, general-purpose plant food formulated for a broad range of foliage and flowering plants. The second is designed for acid-loving species and the third is a safe, gentle organic food which will almost never "burn" delicate roots. All are concentrates which must be dissolved in water and all contain chelated iron, a chlorosis preventive. Stern's two fertilizers also contain important trace elements in addition to the three major nutritional elements, but both must be diluted properly because of their high-nitrogen content.

Finally, never feed a plant that is sick or that is about to be transplanted or has recently been. An ailing plant will deteriorate even more if you add fertilizer shock to its other burdens. Transplanted plants are ready to collapse at the slightest provocation. They are already in mild shock from being uprooted and a dose of high-nitrogen fertilizer might just be the excuse they're looking for to get even with you.

Chapter 3
plant diseases, causes and cures

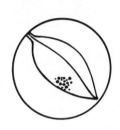

I t may seem at first glance that dealing successfully with diseases which afflict plants requires a PhD in plant pathology or at least a thorough knowledge of botany. *Au contraire.*

In the first place, there aren't that many diseases to which houseplants are vulnerable because of the sheltered environment in which they live. In the second place, good horticultural care will prevent most of these diseases. Still, you should be knowledgeable about diseases and their symptoms so that if your collection is ever stricken by some mysterious ailment, you'll be able to diagnose the illness and treat it successfully.

The following few paragraphs may appear a trifle technical, but they aren't really. Once you've familiarized yourself with the few scientific names of ailments and symptoms, that's just about the extent of the technical terminology.

First of all, let's define what plant disease is: it is anything that causes plants to deteriorate, other than the damage inflicted by predacious insects (aphids, mealybugs, spider mites, etc.). Some diseases are the product of *parasitic organisms* such as fungi, bacteria and virus; others are *physiological* in nature and are induced by environmental conditions: compacted soil, insufficient nutrients (or too many), air pollution and excessive heat or cold, for example.

There are a few terms with which you should be familiar so that you can discuss plant disease problems intelligently with nursery workers and experimental station staff, or so that you can do your own research at the library. Visible signs of plant disease are evidence of the *pathogen* (disease-producing organism). *Necrotic* symptoms are areas of dead foliage and bloom where cells have been destroyed. *Hypoplastic* symptoms are underdevelopment, as in the case of stunting. *Hyperplastic* symptoms are overdevelopment of tissues, as in the case of tumefaction, or galls, abnormal foliage curling, and callus, in which tissue overgrowth is produced in response to a cut or wound inflicted on a plant or tree.

Fungi and Fungus Diseases.

Parasitic fungi are microscopic plants which are devoid of chlorophyll. Since all plants require chlorophyll to produce food, those which don't have it must obtain food from other sources. Parasitic fungi simply attach themselves to higher plant forms and syphon off vitality from their victims.

Fungi attack host plants by inserting suckers (haustoria) into the foliage and removing nourishment from the cells. Fungi soon progress to the reproductive stage during which they produce hundreds of spores. These spores are comparable to seeds, and if carried by wind or on clothing and hands to other plants they spread the fungi to healthy plants. Even though a plant may be otherwise strong and vigorous, these spores may germinate and penetrate the tissues of the plant to devitalize and eventually kill it.

Following are some of the major fungus diseases which attack houseplants:

Anthracnose (an · THRAK′nose)

This disease is more a threat in warm, humid greenhouses but it can occur in any environment of moist, stagnant air. Symptoms are sunken depressions or spots on foliage with gray-to-brown, dry cores and dark or reddish margins.

Controls: Damaged leaves, of course, can't be restored and should be pruned off and discarded in the garbage (not the compost pile). Any of the popular fungicides such as Ferbam, Maneb or Zineb can be applied to the rest of the foliage to protect it from attack. Plants that have once been invaded by anthracnose should be kept on the dry side with a minimum of misting.

Botrytis (BO′trih · tis) Also called "Gray Mold fungus."

Botrytis is responsible for several blights which attack foliage plants and flowers, indoors and out. Foliage and blossoms are covered with grayish mold which can lead to rot. The mold is produced by spores growing at the end of stalks in the diseased areas of the host plant. These can be seen clearly with a strong magnifying glass and explain the Greek name *botrys*, which means a grapelike cluster. Like many fungi, botrytis may produce a black resting body *(sclerotium)* which provides a "nursery" for new fungus production. The *sclerotium* makes it possible for the fungus to over-winter safely through the coldest weather and resume spore production when warm temperatures and humid air return. The *sclerotium* can also live in the debris of fallen leaves and even in the soil around the diseased plant. A good example of botrytis is the grayish mold on strawberries that have gone soft.

Controls: Poor air circulation, moisture left overnight on foliage and flowers, and overcrowding all provide ideal conditions for the development of botrytis. Since it is easily spread by pests (even flies) and transmitted by handling healthy plants after touching infested specimens, fast action should be taken once it is diagnosed. Seriously diseased plants should be discarded. The risk of spreading the disease is too great. Diseased portions of plants which have the disease in its incipient stages should be cut out and destroyed. Zineb, Ferbam and Thiram are three fungicides which can be used with good success in most cases. And two newly developed systemic fungicides, Benlate and Mertect, are also proving to be even more effective. Good ventilation on humid days, less frequent misting and cutting back on the volume of water helps eliminate the source of this disease.

Crown/Stem Rot.

Another fungus disease almost always brought on by overly moist soil conditions and excessive humidity combined with inadequate ventilation. The fungus *Sclerotium delphinii* is responsible for the damage which starts out either by wilting and yellowing tender new growth or rotting the crown and killing the plant, without any preliminaries. Usually the stem is also affected, turning mushy where the tissues have been destroyed.

Controls: Since the *sclerotia* (resting bodies of the fungus) are able to remain dormant but alive in the soil for years, pots, soil and victim should all be discarded. You *can* try to salvage the plant by surgically removing decayed tissue (see the chapter on surgery) and treating the area with Captan or Ferbam, but the wisest move is to sacrifice the plant unless it is extremely rare, valuable or has overwhelming sentimental value.

Damp-off Also called "foot rot."

Anyone who has grown plants or flowers from seed has probably experienced this frustrating, even infuriating, disappointment. Seedlings that have, after weeks of impatient watching, emerged miraculously from the soil suddenly keel over at the soil line and wither. This is the result of damp-off disease and is the foul work of a fungus that is usually present in all but treated soil. The disease may be encouraged by sowing seeds too thickly or keeping the soil too wet and boggy. Nothing can be done to rejuvenate seedlings that have collapsed, but preventive measures against a repetition the next time you sow can be taken.

Controls: The best way to avoid damp-off disease is to sow the seeds in milled sphagnum moss or vermiculite, or in a combination of both. If you prefer to sow seeds in soil, treat it with Panodrench, PCNB (Terraclor) or Semesan, which kill the fungus but not beneficial soil organisms which enhance soil fertility. For double protection, you can also dust seeds before sowing with a fungicide.

Leaf Spot.

This general term covers a variety of symptoms which may indicate that a plant is suffering from either a fungal or a bacterial disease. When large areas of the foliage are covered with dead areas, the disease is called anthracnose or blight or blotch. Any tiny black specks that can be seen on the affected areas are the fruiting bodies of a fungus disease and contain spores. The correct diagnosis, then, is a fungal infection. If no specks can be seen, the disease is probably bacterial. Leaf spot has several causes: humid, stagnant air, overwatering and misting too late in the day so that foliage is still wet after sunset, or a combination of these, plus inadequate ventilation.

Controls: Diseased leaves should be pruned off and destroyed. Then spray the rest of the plant with a fungicide, such as Folpet or Maneb. Improve ventilation on humid days and cut back on irrigation and misting.

Powdery Mildews.

Like most plant diseases, powdery mildews are encouraged to develop by humid, stagnant air. This group of fungi can stunt and dwarf host plants but seldom affects them so severely that they can't be salvaged. Evidence of the disease is seen as a white, powderlike patina on foliage and blooms. This coating is composed of a series of threads (mycelium) and a series of spores (conidia). Feeding is accomplished by root-like organs (haustoria) which pierce the epidermis of the host and feed on the plant cells. Spores are released periodically from the fungus growth and are carried through the air to attack nearby plants. Overfeeding with high-nitrogen fertilizer stimulates the development of mildews in plants.

Controls: This disease can be effectively dealt with by employing such spray fungicides as Acidione PM, Karathane and Folpet (Phaltan). On humid days keep the air moving around plants to avoid the development and spread of the disease.

Root Rot.

Roots that are kept permanently wet will eventually decay. There are no two ways about it. If the topsoil is swampy, the stem may also rot, although it has a better chance since it is exposed to the air. The first sign of root rot is usually the sudden death of seemingly healthy foliage which turns brown or black and curls up. Or the entire plant may droop and wilt. The soil develops a pungent sour smell which is created by decaying root tissue.

Controls: Treatment techniques for this condition are described in the following chapter.

Sooty Mold.

This black fungus appears on plants which have serious infestations of scales and aphids. Although it is unsightly, it is not directly harmful to plants since it isn't parasitic. Its total diet is the excrement known as "honeydew" produced by insect predators. Sooty mold is a combination of black mycelial threads and spores. Its potential for plant damage comes when it has covered the epidermis of foliage to such an extent that light is blocked out. Plants must have light for photosynthesis, the manufacture of food.

Controls: With the eradication of pests and honeydew comes the elimination of sooty mold. When its food supply disappears, it does, too. It's as simple as that. For control of honeydew-excreting pests, use a systemic pesticide such as Isotox granules in the pot soil.

Bacterial and Viral Diseases.

Parasitic bacteria are microscopic, single-celled plants which multiply by fission. Fission, you'll recall from your high school biology, is the reproduction of an organism by division. Bacteria are responsible for a number of plant disorders—blights, rots, wilts and leaf spot, to name a few. Like fungi, parasitic bacteria invade plant cells and eventually destroy plant tissue. Although some forms produce resting bodies, or spores, which enable them to exist in plant debris and dormant bulbs, seeds, etc., most are unable to create spores by which to spread their infection to healthy plants through the air.

Viruses are even smaller than bacteria and can only be seen with an electron microscope. Symptoms of viral disease in plants are varied. The disease may manifest itself by stunting, distortion and curling of foliage which is accented with irregular green or yellow spots and color breaking of blossoms in which white streaks appear in colored petals.

Both bacteria and viruses can invade plants through pores (stomata), cuts made by pruning shears or through the holes inflicted by predacious insects. Insects often act as virus vectors, injecting the disease picked up from an afflicted plant into a healthy specimen while they feed.

Plants, unlike man and animals, are unable to produce antibodies to fight bacterial and viral diseases and there are no known reliable cures. There are bactericides which *prevent* the invasions of the diseases, but they are ineffective in curbing them once they've invaded a plant. Both the plant and the potting soil (and the pots as well if they are clay) should be discarded in the garbage can. Viruses in particular are capable of surviving for some time in soil, ready to attack anything planted in it.

Physiological Plant Diseases and Problems.

More plants deteriorate and die from environmental or physiological problems than all other disorders and pest damage combined. Nine times out of ten, these problems are the result of our neglect, indifference or ignorance and could have been prevented if we had been more alert and attentive. Some, of course, we have no direct control over (air pollution, for example), but even these can be ameliorated or overcome with a little extra effort.

Diet excesses or deficiencies.

At one time or another in our avocation of collecting and cultivating plants, we've all probably been guilty of overfeeding them. Our motives are usually commendable. We like to pamper our plants. Our indulgent mothers kept us stuffed with goodies as one way of demonstrating their affection for us, so we stuff our plants with food to display our affection for them. Or maybe somewhere we read or heard that by doubling up on plant food we could induce our plants to grow bigger and better. If a little food is good, isn't a lot of food that much better? If plants could talk, their answer would be a resounding "no!"

In a way, plants *do* communicate. Their response to overfeeding is demonstrated physically in a number of ways. The first indication that plants are being given too much or too strong a dose of fertilizer is "burning" or browning of tender

Excessive fertilizer causes tips, margins of leaves to brown off.

leaf tips and margins. Total collapse of the plant may occur if the nitrogen content of the plant food is too high.

Just as excessive amounts of fertilizer are harmful to plants, a dearth of adequate, balanced nutrients is equally damaging to their physiology. Plants need more than soil, water and light to survive and thrive. If they were growing outdoors in the soil, they would have access to all of the elements they require, since most of these exist naturally in the soil. But when we stick our plants in pastuerized soil and confine their feeder roots to the limited space of a container, they are totally dependent upon us to provide the occasional boosters of nutrients they need.

As we've seen in the chapter on basic plant care, plants require a complete fertilizer—one that contains nitrogen, phosphorus and potassium, plus the various trace elements which contribute to plant development and health. Following are symptoms of dietary deficiencies in plants:

Boron deficiency. New leaves are the first to exhibit signs of inadequate boron. They often appear pale and yellow-green, and feel brittle. Stems may also be brittle. Highly alkaline soil is usually devoid of sufficient boron.

Copper deficiency. New growth emerges an unnatural dark green or blue-green, but as the leaves mature they change to yellow or pale greenish-yellow.

Iron deficiency. A lack of sufficient iron is probably the most common deficiency in plants. New growth, again, is the first place the problem is recognized. The tips of young leaves turn yellow. Eventually, the entire leaf, except for the network of veins which remain green, also turns yellow. The condition may be induced by a nitrogen deficiency or inadequate oxygen in water-logged soil, but the most common cause is a lack of sufficient iron. Acid-loving plants grown in highly alkaline soil often develop the condition which is technically termed chlorosis. Fortunately, nine times out of ten this problem can be corrected in a few days by adding chelated iron to the soil. Chelated iron is marketed under a number of different brand names. Chlorosis can be avoided by using a fertilizer that contains iron chelate, not merely iron. A chelating agent makes the element immediately available to a plant's system.

Magnesium deficiency. Blotches of gray-green, yellow-green or light brown on mature leaves are the first symptoms of magnesium deficiency. New foliage emerges abnormally thin and soft and dwarfed. This deficiency is frequently found in plants grown in soil that is too acid and sandy.

Manganese deficiency. Immature leaves exhibit dark green veins while the rest of the leaf is pale green. Soil containing excessive limestone is usually the culprit.

Chlorosis is a condition caused by deficiency of chlorophyll. Entire leaf is yellow, except for midrib and veins.

Nitrogen deficiency. An insufficiency of nitrogen, one of the most important elements, adversely affects plants in several ways. Growth usually stops or slows to the proverbial snail's pace. Leaves may begin to pale, changing from normal green to greenish-yellow to yellow. New leaves never achieve normal size and may exhibit reddish tinges in the veins and in the petiole.

Potassium deficiency. Older, mature leaves are usually the first to show the effects of a lack of sufficient potassium. Leaf margins become auburn or brown at first, then turn orange or purple. Tissues at the tips and margins of leaves may deteriorate and die.

Air Pollution.

An unhappy by-product of big cities and industrialization is smog and other pollutants which are potential killers of human as well as plant life. Smog used to be thought of as a peculiarly California offspring, born in the Los Angeles basin from an unhealthy romance between Angelenos and their automobiles. Well, that just isn't the case any longer. Every major city has smog periodically, depending upon the atmospheric conditions. In New York, for example, scientists have determined that airborne contaminants weighing nearly *fifteen tons* are dropped on every square mile of the city each *month.* Japan holds the record for the worst pollutant fallout: between the cities of Osaka and Tokyo almost thirty tons of crud settle on every square mile each month.

Smog is a combination of all the products of combustion—soot, sulphur dioxide, sulphuric acid, ethylene—and ozone. Obviously, then, auto emissions aren't the only culprit, although nearly seven million tons of lethal nitrogen oxides are pumped into the air every year by automobiles.

During the summer months, when pollution and smog are at their worst, the danger to houseplants is at its peak since windows are left open for ventilation most of the time. Air conditioners virtually eliminate this problem, but few city apartment dwellers enjoy this luxury.

Symptoms of pollution damage are leaf spotting, edge browning and curling. Deformed new growth is symptomatic of severe pollution. Outdoors, the problem is becoming catastrophic in some areas where drifting city and industrial pollution is denuding and destroying entire forests, forcing growers to develop smog-resistant plants, flowers and trees to survive in our poisoned environment. Their success to date has been marginal.

Ideally, the way to cope with pollution damage is an air conditioned environment which eliminates the need to open windows for "fresh" air. Since this is not practical for most people, the next best thing is to wash foliage down once a day during peak smog months with suds from an Ivory soap bar in warm water, followed by a lukewarm rinse of clear water, especially if windows have been left open. (African violets can't be bathed in this manner.) On really bad days, when the air is also muggy, use a fan to keep the air moving and hold off on misting and watering. If you live in an area or city where air pollution is a more or less continuous problem, build a collection of thick-leaf plants, such as rubber trees (*Ficus decora*) and Kentia palms (*Howeia belmoreana*), which can withstand the effects of pollution and subsequent scrubbings far better than tender thin-leaf counterparts, such as delicate ferns.

Tobacco smoke. Let's face it, the tars and nicotine contained in cigar, cigarette and pipe smoke are not *beneficial* to plant life, but neither are they as harmful in moderation as some people would have you believe. The operative phrase is "in moderation," which means that if only one or two members of your family smoke, your plants probably will adjust. Good ventilation and wiping the leaves monthly to remove the characteristic yellow residue to clear the stomata are strongly advised countermeasures. Tobacco smoke *does* become a problem when there is a heavy concentration of it in the air in a closed environment. Parties, where a lot of the celebrants are puffing away, can be disastrous to thin-leaf species, particularly ferns. And, not just from the smoke but from the numerous cigarettes thoughtlessly snuffed out in plant pots by callous visitors (who should be made to remove same with their tongues). When you're having a group over, it might be wise to isolate your sensitive plants in a room where they won't be exposed to smoke.

Gas injury. In the days when manufactured gas was used in cooking and illumination, gas injury from leaks killed many a Victorian plant. Natural gas is dangerous to most plants only in great volume—in a closed room, for example. Leaf spotting, wilt and collapse are the symptoms of gas injury.

Paint fumes. Fumes from most paints, varnishes, etc., are as noxious to plants as carbon monoxide is to us humans. When you paint, plants should be moved out of the room, not simply covered with a drop cloth. It is safe to bring plants into a newly painted environment only when you find the air breathable again.

Oedema.

Symptoms of this condition are small round holes that appear in leaves, or waterlogged areas in the foliage that turn reddish-brown after a few days. These are the result of blisters that have erupted. The condition results from excessive water intake by the plant and poor or inadequate transpiration. The cause is a combination of warm, moist soil and cool, humid air. Irrigating plants indoors on cool, overcast days creates the perfect conditions for oedema.

Oedema damage. Holes are created by ruptured water blisters.

Fortunately, oedema is controllable if certain steps are taken. Move afflicted plants to a warmer location and allow the soil to dry out. Don't mist until the plant is fully recovered, which may take several weeks.

Pot-bound Plants.

A thriving plant usually outgrows its container in about two years. Its root system will need more room to expand and a greater volume of soil from which to collect food after this time or the top growth will begin to deteriorate. In a chronic case, the plant's roots actually push the crown up above the rim of the pot and roots begin to creep out of the drainage hole in their search for more growing space.

Once a plant's roots have completely filled a container, water and fertilizer pass through the pot like a dose of salts. There is very little soil left to trap and hold moisture and nutrients until the roots can absorb them. Soon, the feeder roots—unable to find nourishment—begin to dry out and shrivel. When things reach this stage death is imminent.

Nearly all plants when they're young benefit from being slightly pot-bound. When the roots are cramped, energy is transferred from the production of an intricate root system to forming more foliage and flowers. The result is a fuller, more attractive plant. But this condition should not be allowed to get to the point where the plant is on the verge of collapse from starvation and root shock.

Plants should be knocked out of their pots every spring for a root check. If they have wrapped themselves completely around the outside of the soil ball and very little else is visible, it's time to move them up to pots an inch or two larger. If the roots still appear to have ample soil left to fill, slip the pots back on, tap the bases on a solid surface to reset them, and wait until the following spring.

When you do pot on (move up in pot size), use one only one or two inches larger. If you "jump" the plant (pot on in a container too large for the root bulk), the plant will spend the next year merrily filling the container with roots and you'll see almost no top growth.

If you have a pot-bound plant but you want to keep it in its original pot because it fits perfectly in a decorative container, this can be accomplished, if you're adventurous. It involves giving the root ball a slight haircut. Care must be taken not to cut too much of the root structure away, particularly near the central root system. Remember, you're giving it a trim, not a butch. Dip the remaining roots in a rooting hormone and repot. Moisten slightly, but don't soak; then let the soil dry out for two

Plants left too long in original container become seriously pot-bound.

weeks. The drying of the soil and the wounding of the roots will stimulate the development of new roots.

When you repot, make sure the soil is packed firmly (but gently) around the roots. A loose pack leaves airpockets. Roots extend themselves in response to the pressure of soil against them. If the soil is not pressing tightly against them they won't penetrate it. This means the plant will not continue to thrive, although it may *survive* indefinitely.

One final word of caution: never fertilize a plant before or just after transplanting or repotting. The roots are already traumatized by being uprooted and fertilizer only adds to the shock.

Salts Injury.

Fertilizer salts build-up in pots is a common problem for those gardeners who overfeed their plants, and the problem is compounded when the plants being overfed are potted up in plastic or dry-welled (no drainage hole) containers. The fertilizer given to a plant is never totally absorbed. There is always a residue which remains in the soil. The more one overdoses a plant or the older the soil in a pot, the greater the concentration of undissolved mineral salts. Clay containers with drainage holes postpone the effects of this build-up since some of the salts are able to leach through the porous sides of the container and a little more is flushed out the drainage hole. Often the damage is done to delicate feeder roots, which are burned so severely they can no longer function.

White crusts on clay pots and topsoil are symptoms of too much fertilizer.

Other symptoms of salts injury are browned-off leaf tips and margins, and new foliage that emerges distorted or stunted. The white, crusty build-up of salts on the topsoil, which is typical with this condition, can eat right through the stem of a plant like corrosive acid, and—if there is a layer of salts on the rim—any leaf resting there for long will be severely burned.

Once the build-up of salts has reached the point where foliage is deteriorating, attempting to flush the concentration out of the pot with water is no longer productive. The soil is probably in such a state that only repotting in fresh, virgin soil will save the plant from further damage or death. Pots encrusted with salts and the green scum (algae) which grows on them can be scrubbed clean using steel wool or a wire brush and a mixture of TSP and scouring powder. Rinse several times in clear water to remove all chemical residue. Rims of pots can be coated with melted wax which not only protects leaves from salts injury but, also, against the damage inflicted by the hard, abrasive surface of the clay.

Shock.

Three kinds of shock, or trauma, adversely affect plants, causing them to collapse or "throw" their foliage.

One form of trauma is brought on by the plant's exposure to cold drafts. This is a common result of placing a plant in an entry hall near a door that is opened and closed often in winter, or situated too near an air conditioner where it gets the full blasts of chilled air.

Another kind of shock that occurs frequently at the hands of the inexperienced gardener is root shock brought about by irrigating a plant with water that is too hot or too cold, or inflicted when a plant is potted up. In the first instance, all water poured into a pot should be tepid, or room temperature. Some well-meaning plant experts have gone on record saying that the temperature of irrigation water really makes no difference since water is quickly brought to a safe temperature level as soon as it hits the potting soil. That sounds scientifically plausible, but it isn't the case, particularly on cool days when the soil is already cold and then frigid water is poured into the pot. Be on the safe side, for your plant's sake. Give it tepid water.

Root shock from potting up is an inconstant thing. It is one of those frustrating experiences because it usually occurs after you've been extra careful to avoid it. In most cases, root shock is inflicted when the potting soil isn't wet enough before you try to knock the plant from its container. Feeder roots that have grown into the dried soil caked firmly on the interior walls of the container are rudely ripped away when the plant is removed. The plant recoils in shock and usually collapses.

Occasionally, a root-shocked plant can be saved by trimming away some of the root structure and some of the top growth. Don't cut into the central root; trim away around the edges. Since plants have a mysterious and wondrous will to survive, this wounding of the root structure usually stimulates a plant to reproduce new feeder roots quickly to save itself. Pruning away top growth, while it sometimes detracts from the appearance of the plant, gives the roots a fifty-fifty chance, since they don't have to work so hard manufacturing food to support heavy foliage while struggling back from death's door. It also helps the plant retain more vital moisture which would have been lost by transpiration through the foliage that was removed.

Finally, there is a form of shock which nearly everyone who collects plants has experienced—the sudden collapse of a plant recently acquired or one moved to a new location within the home. Plants sometimes resent being moved. When they are relocated, the obstinate ones show their displeasure by dropping foliage, going into (synthetic) dormancy, or collapsing altogether. It doesn't happen all the time, but when it does it's a real puzzler.

Prevention of this kind of trauma is far from reliable or exact. Newly acquired plants should be given environments that duplicate as closely as possible the ones in which they were thriving before you adopted them. This should prevent relocation shock. Plants received as gifts are another matter. Your best bet is to research their requirements in a horticultural reference.

Solutions to relocation shock are not terribly scientific. A plant that has been traumatized by being moved within the home will probably adjust to its new location if conditions (light, temperature, humidity, etc.) are similar to what it was getting. If they are drastically different, you may be asking too much of the plant—but who knows? After a period of settling in, the plant may recover and begin to thrive. Don't wait until the plant has seriously deteriorated, though, to decide it's not going to see things your way. A period of two weeks should be enough to reveal the plant's attitude to its new location. If it hasn't perked up by then, give in and return it to its original location.

New arrivals are notorious for turning up their noses at new surroundings, particularly if they've spent most of their lives in a temperature- and humidity-controlled greenhouse. From their point of view, your home is too hot or too dry or too light or too dark. Like prima donnas, they resent being trundled about like so much stage scenery. Usually, new plants are young enough to take a lot of abuse in stride, after a brief period of pouting and sulking. They may drop some foliage or droop a little, but in most cases good, attentive care will bring them around eventually. Your surest method of revitalizing them is to withhold water for a few days, provide good humidity and light, and think twice before saying nasty things around them.

Soil Problems and Solutions.

Only after problems involving soil deficiencies crop up do many novice gardeners realize that soil is not just dirt and that not just any old potful will do for growing plants. Plants, of course, *will* survive in virtually any soil, but not without some deterioration. Here's why.

For the cultivation of trouble-free plants, the soil must have fifteen elements, all in proper balance. These elements are carbon, hydrogen, oxygen, nitrogen, phosphorus, potassium, calcium, copper, magnesium, manganese, boron, sulfur, iron, zinc and molybdenum. If any one of these elements is absent, the deficiency will be reflected in the plant.

Each element performs specific functions in plant growth. Calcium promotes the development of plant cells and helps in the absorption of nitrogen, phosphorus and potassium. Iron is needed to help make chlorophyll. Magnesium is the key element in chlorophyll and aids in the distribution of phosphorus throughout the plant. Phosphorus is essential in the formation of a plant's root system, hastens maturity and spurs the production of buds and flowers. Nitrogen is essential for the development of healthy green leaves. Potassium contributes to the production of sugars and starches and improves a plant's capacity to resist disease. The other nine elements help carry on the various chemical processes in plant physiology.

The second consideration in determining soil quality and its suitability is its pH analysis and whether the plants you are growing require an acid or an alkaline medium. This may appear a mite technical, but it isn't really. A scale was developed to measure acid/alkaline values of soil. It's called a pH scale. A pH value of 7 is neutral. Anything from 7 up to 14 is alkaline; anything from 7 to 0 is acid. The value increases ten times with each unit up the scale and, conversely, decreases ten times down the scale. You can buy a soil-testing kit at garden centers or hardware stores for under $10 which will enable you to analyze your potting mixes for the presence of the fifteen elements and the pH value. It's all duck soup.

A less-exact method of testing soil for acidity/alkalinity is one that has been used for decades—the litmus paper test. Strips of red and blue litmus paper are pressed against damp soil samples for a few minutes. If the red paper turns blue, the sample is alkaline. If the blue paper turns red, the soil is acid. The soil is neutral if both strips turn purple.

Garden loam and soil from the yard or woods is not suitable for house plants. It carries larva of destructive insects and, sometimes, the bugs themselves; it is probably riddled with fungi and harmful bacteria; it is usually loaded with harsh fertilizers that are too strong for houseplants; and it is probably deficient in many of the elements indoor plants require.

When you need soil for your houseplants, buy the packaged pasteurized mixes available wherever plants are sold. You can also buy the various ingredients and amendments and mix up your own formulas to suit the needs of different species.

One of the commonest problems indoor gardeners encounter is compacted soil—hardpan is the technical term. The soil particles are so finely ground down and tightly packed together that water takes forever to penetrate and air, carrying the life-sustaining oxygen plants need, doesn't even try to get through the cementlike composition to the roots.

Hardpan is usually the result of years of overhead watering. The soil particles are eroded more and more and packed tighter and tighter each year from weekly irrigations until, finally, you need a jackhammer to break through the topsoil.

This condition can be avoided by repotting plants every three or four years in virgin soil. By this time, the soil in a pot is in pretty bad shape anyway and has lost most of its fertility and friability (its ability to be crumbled into fine particles). It usually has collected a dangerous concentration of undissolved mineral salts, too, so repotting in new soil prevents other problems later on.

Diseases and their Prime Targets

Anthracnose

Ferbam, Maneb, Zineb

- Aglaonema   *front up ½*
- Avocado
- Ficus
- Kalanchoe
- Palm
- Privet

Root Rot

- African violet
- Aloe
- Begonia
- Cactus
- Gloxinia
- Kalanchoe
- Palm
- *Creeping Charlie*

Crown Rot

- African violet
- Aglaonema
- Begonia
- Cactus
- Delphinium
- Geranium
- Gloxinia
- Philodendron

Stem Rot

- African violet
- Begonia
- Cactus
- Dieffenbachia
- Geranium
- Gloxinia
- Palm
- Philodendron

Diagnosing
Plant Diseases and Ailments

Condition	Probable Causes	Remedies
Foliage Plants		
	Excessive heat	Improve ventilation or lower thermostat.
	Inadequate humidity	Mist frequently.
Brown, dry leaf tips and margins. Plant foliage feels leathery.	Overwatering	Let soil dry out. Cut back on water volume.
	Salts injury, fertilizer build-up in soil	Flush excess fertilizer out of pot with water. Cut back on frequency and amount of plant food.
Plant collapse. Total wilt.	Root rot	No sure cure.
	Shock from excessive cold, heat, repotting trauma	No sure cure. Plant may be saved by root/foliage trimming. See Chapter 5.
	Fertilizer "burn"	Flush out excess fertilizer with water. Repot in fresh soil. No 100% sure cure.
	Dehydration	Deep water and mist.
Foliage limp but green.	Excessive heat	Improve ventilation. Lower thermostat.
	Dehydration	Deep water and mist.
	Pot-bound	Repot plant in larger container.
Leaf drop	Overwatering	Let soil dry out. Cut back on frequency and volume of water.
	Cold drafts	Move plant away from source.
	Cold water	Bring irrigation water to room temperature before pouring in pot.
	Gas injury	Either move plant to another location or have source of leak corrected.

Condition	Probable Causes	Remedies
New growth emerges yellow.	Iron deficiency	Irrigate plant with solution of one ounce iron sulfate in two gallons of water. Repeat every two weeks until green color is restored.
Paling (etiolation) of foliage.	Insufficient light	Move plant to location that gets more intense illumination or supplement with artificial light.
White or yellowish rings or spots on foliage.	Irrigating water too cold. Watering tropicals with frigid water also creates danger of shock.	Bring irrigation water to room temperature before using.
Slow growth.	Inadequate light	Move plant to location that gets more intense natural or artificial light.
	Nutritional deficiency	Give plant complete fertilizer monthly from spring to fall.
Depresions or spots on foliage with greyish-brown dry cores and dark or reddish margins.	Anthracnose disease	Treat with fungicide. Keep plant on the dry side with minimum misting.
New growth rapid but weak.	Excessive fertilizer	Flush out excess fertilizer with water. Withhold all food for three months to allow plant to bounce back.
Leaves gray or watery in spots.	Bacterial blight	See earlier in chapter for diagnosis and control of specific blights.
Leaves coated with white, powdery substance.	Powdery mildew	Dust with Zineb or sulfur. Improve air circulation.
Leaves develop yellow or brown spots, some curling.	Sunburn	Filter direct sun or move plant back from direct solar rays.
	Air pollution	Wash foliage with soap and water. Keep air moving to prevent settling of pollutants on leaves.
Gray mold on foliage. Spores visible under magnifying glass.	Botrytis cinera blight	See earlier in chapter for specific treatment.

Condition	Probable Causes	Remedies
Leaves widely spaced. Growth thin and weak. New leaves small at maturity.	Insufficient light	Move plant closer to light source. Supplement natural with artificial light.
Plant rots just above soil line.	Crown rot	No cure.
Seedlings wilt and die soon after sprouting.	Damp-off fungus disease	No cure for dead seedlings. In future, sow seeds in milled sphagnum moss.
Black fungus appears on stem and foliage.	Sooty mold	Eliminate pests on plants which excrete the honeydew on which sooty mold grows.
Small round holes and blisters appear in foliage.	Oedema	Move plant to warmer location. Allow soil to dry out. Hold off on misting until plant is fully recovered.
Flowers fail to appear.	Insufficient sun, light	Nearly all flowering plants need four to five hours of sun a day.
	Inadequate moisture Excessive fertilizer	Keep soil evenly moist. High-nitrogen fertilizer should be withheld at the time of year plant normally blooms.
Buds drop.	Rapid temperature shift, drafts	Keep plants in draft-free, even-temperature environment unless specific culture requirements say differently.
	Inadequate humidity	Provide proper humidity level required by misting or creation of micro-climate.
Gray mold on flowers. Spores visible under magnifying glass.	Botrytis cinera blight	See earlier in chapter for treatment recommendations.
White spots on Azalea blooms, Camellias turn brown.	Blights	See earlier in chapter for treatment recommendations.

Flowering Plants

Chapter 4

predacious pests, their damage and control

Some indoor gardeners, discovering a single insect on one of their plants, launch a chemical assault on the luckless host plant capable of wiping out every mosquito from here to Malaya. While their fast action is commendable, their overkill approach is an overreaction to a common occurrence. As long as you have plants, you're going to have bugs that want to chomp on them. It's no crime if a critter has outwitted you and sneaked in to prey on your collection, but it *is* criminal to allow a colony of destructive predators to establish itself before you act.

Regardless of how circumspect you are in selecting new plants, no matter how carefully you examine plants set out to air before bringing them back inside, or how many cracks and crevices you've sealed to keep predators out, some will get through your defenses. They may fly in unnoticed in the moment it takes to open or close a door. Aphids don't even need that much room. Most can wriggle through even fine-mesh screens. So accept the fact that if you want plants, you'll have bugs, too, and be prepared to cope with them.

ANTS

Description/Habits.

Ants are more an unsightly nuisance than they are pests. Ants in your plants are almost always a sign that you have a pest problem. They are attracted to plants infested with aphids, scales or whiteflies by the honeydew these pests excrete. The ants gather honeydew for food and, clever devils that they are, often "farm" aphids by carrying them from plant to plant to establish new colonies to increase the availability of honeydew.

Common Ant

Damage.

Sometimes the burrowing of ants in the pot soil can damage young plants and seedlings by disturbing roots. The more serious threat is that ants spread the infestation of plant predators by carrying them to healthy specimens.

Controls.

Ant baits which are usually available at garden centers, hardware stores and even supermarkets, are effective. An age-old method of organic control is to sprinkle dried tansy leaves around points of entry or exit of ant columns. An obvious solution is to eradicate the predators which excrete the honeydew which attracts them initially.

APHIDS.   Also called "plant lice."

Description/Habits.

Aphids probably hold the record·for instant population explosions. One day a plant may be "clean" and the next day literally covered with active dots which may be green, yellow, brown, red or black. Aphids have soft, pear-shaped bodies and long legs. Some are winged, others are wingless. Nature has programmed these minute insects to survive by equipping them with special talents. Young aphids, or nymphs, grow rapidly and become "stem mothers," giving birth to living young without mating and by by-passing the normal egg-laying/incubation cycle. In just one week, these infants repeat the process and in short order hundreds of gluttonous para-sitic mouths are sucking away at the plant's lifeline. But that isn't the end of the bad news. Some of the females sprout wings and move on to establish new colonies on other plants and the entire procedure is repeated *ad infinitum*. To top it all off, aphids gum up stem and foliage with their sticky excrement, euphemistically called honeydew.

Damage.

Aphids feed by piercing the epidermis of a leaf with their sharp stylets and sucking sap from the plant cells. Because they are so prolific, a plant can be seriously weakened in just a few days. Aphid damage is usually first noticed on the tender growing tips of leaves, where these pests prefer to colonize. Stunting, distortion and discoloration result from their feeding

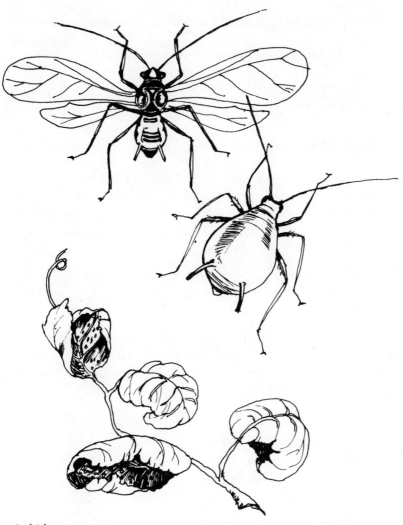

Aphids

activity. Flowering plants afflicted with an aphid invasion produce bizarre, deformed buds and flowers of pale, unnatural color. Some aphids feed on roots and these insidious beasties are almost never discovered before they've done their dirty work and killed the host plant.

Favorite Targets:

Aphids are omnivorous. Almost every species is threatened by them.

Control:

Aphids are among the easiest of pests to control. Usually, a strong spray of lukewarm water will dislodge most, but spraying the afflicted plant with a solution of warm soapy water and Malathion is a sure cure. Root aphids must be attacked with a systemic or Malathion drench poured in the soil.

BROAD MITES

Description/Habits.

Similar to cyclamen mites (discussion below), broad mites are nearly transparent and a bit smaller. Unlike cyclamen mites, who like to invade buds and deep folds of foliage, broad mites congregate on the underside of leaves, making them more vulnerable to chemical assaults.

Damage.

Broad mites cause bloom and foliar distortion similar to that of cyclamen mites, but broad mites leave a shiny, silvery cast on deformed areas where they've been feeding. Deformed leaves may also become brittle and hard.

Favorite Targets:

African violet, cyclamen, geranium.

Controls.

Sulfur dust applied at weekly intervals for three weeks usually assures control of broad mites, although sulfur is largely ineffective against cyclamen mites. Cedoflora applied every ten days for a month works well, also, as does Dimite.

COCKROACHES

Description/Habits.

As unlikely as it may seem, cockroaches will attack houseplants, particularly if their ordinary food sources have been eliminated. They are attracted to plants already deteriorating from fungus disease or plants whose topsoil is covered with decaying leaves and other debris where they can hide.

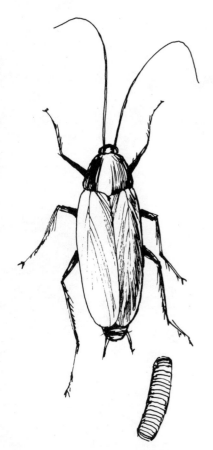

Common Cockroach

Damage.

Roaches gnaw and chew at tender stems and leaves, leaving ragged leaf edges and scarred stems.

Favorite Targets:

New, tender growth of foliage plants.

Control.

Roaches are difficult to control once they've gained a foothold, particularly in older city apartment buildings where one tenant's attack on them only drives them into another tenant's abode, setting up a never-ending cycle of infestation and reinfestation. Plants can be protected to some extent by spraying Malathion on the foliage. Keep topsoil clear of decaying leaves and other debris.

CRICKETS

Description/Habits.

Crickets are related to grasshoppers and have very effective, destructive chewing mechanisms. They favor dark, damp areas and hide during the day, making detection difficult. Terrariums are particularly favored hiding places and just one cricket can ravage a terrarium or bottle garden in one night. If conditions are ideal, it is not uncommon for crickets to lay eggs indoors, posing a real threat to a plant collection.

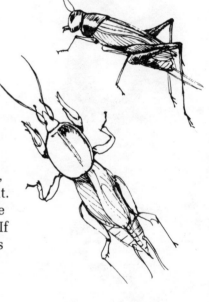

Common Cricket (top),
Mole Cricket (bottom)

Damage.

The overnight disappearance of new leaves and tender young shoots is usually the work of crickets.

Favorite Targets:

Seedlings and tender new growth.

Controls.

Malathion as a drench or spray is effective. Other controls are nicotine sulfate and systemic pesticides, such as Isotox.

CYCLAMEN MITES

Description/Habits.

Cyclamen mites are probably the most difficult of all houseplant predators to eradicate. They are almost microscopic—1/64th to 1/100th of an inch long—and their tiny size enables them to crawl into the almost inaccessible folds and crevices of flower buds and leaf folds where they usually successfully escape sprays and other control attempts. Although they are named for the host plant they prefer, African violets are on their menu along with dozens of other popular flowering and foliage houseplants. Like aphids, they congregate at the tender growing tips of leaves and inside flower buds where there is plenty of sap to support their insatiable appetites. A cool, damp environment is ideal for rapid colonization.

Damage.

Every part of a plant is affected by cyclamen mite infestation. Buds that open produce misshapen, discolored, blotchy flowers and shriveled, weirdly shaped foliage with a scabrous, lumpy surface. Leaf margins are sometimes turned up and the entire leaf may look pale greenish-yellow. Both buds and foliage have a tendency to drop at the slightest touch. Often, cyclamen mites serve as vectors for bacterial and fungus infections and chlorosis is not uncommon in badly infested specimens.

Favorite Targets:

Cyclamen, African violet

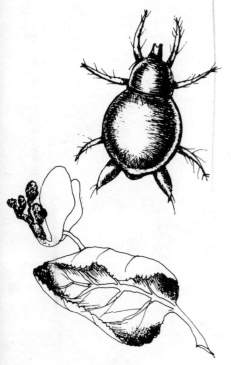

Cyclamen Mite

Controls.

Because cyclamen mites are so difficult to eradicate and since they can proliferate rapidly to infest other plants, the wisest move is to discard host plants in the garbage can. Before pesticides were developed to deal with them, the control technique used, with occasional success, was to submerge the plant in water kept constantly at 110°F for fifteen minutes. (Don't use this method on African violets.) A miticide, such as Dimite, has proved somewhat successful and an organic compound, Cedoflora, is a pleasant-smelling treatment that is moderately successful if sprayed weekly for at least a month.

EARWIGS.   Also called "European earwig."

Description/Habits.

Earwigs have a worse reputation than they deserve, probably because of their sinister appearance and the widespread Old World myth that they crawl into the ear of a sleeping person and imbed themselves in his brain. They are a reddish-brown color and are seldom larger than an inch. Males have large, curved forcep-like appendages extending from the abdomen; females have smaller, straighter pincers. Females lay eggs in the spring and, unlike most insects, watch over them until they hatch and molt. Earwigs are night-feeders and prefer dark, damp places and can usually be found under flower and plant pots and garden debris. They are usually brought indoors in the blooms and buds of cut flowers from the garden and in houseplant pots set outside to air. The European earwig was first seen in the United States in Rhode Island in 1911. Just a few years later, like the common garden snail imported from France, they had spread to the West Coast.

Damage.

Earwigs chew holes in foliage and blooms. Young earwigs prefer tender young shoots but will also nibble on blossoms and vegetables.

Favorite Targets:

Omnivorous, but a preference is shown for new growth and herbaceous plants.

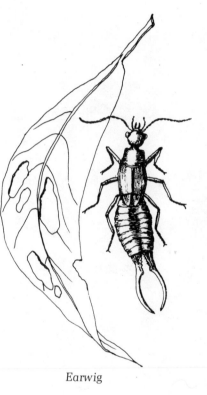

Earwig

Controls.

If you suspect that earwigs are responsible for holes in foliage and flowers that mysteriously appear overnight, submerge the plant, pot and all, in warm water for a few minutes. If there are any earwigs nestled in buds, flowers, foliage or pot, they'll abandon ship since they loathe water. Any discovered swimming about can be removed with a spoon and consigned to a jar of soapy water. It really isn't necessary to take stronger measures than the pot submersion technique, but if there is some reason why this is impractical, use a systemic pesticide in the soil and handpick (with tweezers) any earwigs found hiding inside flowers and buds.

FUNGUS GNATS

Description/Habits.

Anyone who has ever owned a houseplant has probably noticed one or two tiny critters flying around a recently watered specimen. These are fungus gnats which seem to multiply the more frequently you water your plants. As far as has been determined, they cause no *direct* damage to plants but their progeny do. Periodically, they lay eggs on the surface of the soil. These eggs hatch into tiny white maggots. The more organic the soil (peat moss, leaf mold), the greater the brood.

Damage.

Some fungus gnat maggots burrow down into the soil and bore their way into roots. Others devour feeder roots, and still others the crowns of plants, inducing rot, weak growth and chlorosis. Scars and holes inflicted on roots open the door for the development of bacterial diseases.

Favorite Targets:

Any moisture-loving plant or one whose topsoil is kept constantly wet.

Controls.

Allow soil to completely dry out on the surface before watering again. This will eliminate fungus gnats. To control maggots, drown them out and bring them to the surface by immersing the pot in a bucket of water and leaving it there overnight. A drench of limewater, which can be obtained at the pharmacy, poured liberally over the topsoil is also effective. Dousing the soil with Malathion kills both eggs and maggots.

LEAFHOPPERS

Description/Habits.

Closely related to aphids, leafhoppers are slender, wedge-shaped, highly energetic insects with two pairs of delicate, lacelike wings which they hold high over their backs when in repose. When disturbed, they fly and hop wildly

Leafhoppers

around and may land on nearby plants to spread the contagion. White, translucent fragments found on foliage and topsoil are the skins of nymphs which have molted.

Damage.

Leafhoppers extract sap from the undersurfaces of foliage and their punctures leave a series of white dots which are visible on the upper leaf surfaces. There is fairly conclusive evidence that leafhoppers inject a toxin into plant cells as they feed which causes leaf margins to curl and brown-off. They are also suspected as vectors of viral and other plant diseases.

Favorite Targets:

Citrus and herbaceous plants.

Controls.

Malathion, used both as a soil drench and spray, is successful against this pest. Also, a systemic like Isotox works well.

LEAF MINERS

Description/Habits.

Leaf miners are the larvae of moths, beetles, maggots, flies, caterpillars, etc., which have hatched between the upper and lower epidermis of a leaf. A number of insects lay eggs inside leaves, providing the larvae which hatch with a ready-made larder of succulent food. Of the several types of leaf miners which prey on plants, trees and flowers, the three most common in the culture of houseplants are the chrysanthemum leaf miner (*Phytomyza chrysanthemi*), or marguerite fly; the palm leaf miner (*Homaledra sabelella*); and the privet leaf miner (*Gracilaria cuculipennella*).

Leaf Miner

Damage.

Deterioration of plants, leaf by leaf, results from the tunneling, or mining, activity of these larvae. Some produce linear mines which are serpentine in configuration, others make blotch mines, which are white, circular ruptures where cell tissues have been eaten away that eventually turn into dead, brown areas on the surface of the leaf.

Favorite Targets:

African violet, ficus species, palms and privets.

Controls.

Since leaf miners burrow inside foliage, sprays are largely ineffective in dealing with them. The only surefire control is the use of a systemic pesticide introduced into the soil. Badly mined leaves, which may still contain active leaf miners, should be pruned from the plant and destroyed.

MEALYBUGS

Description/Habits.

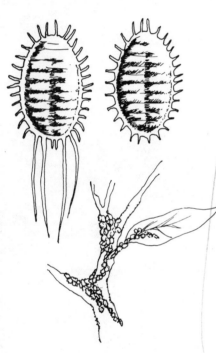

Mealybugs—Long-tailed (left), Short-tailed (right)

Mealybugs are so common a pest that at one time or another most of us have adopted a new plant only to discover when we got it home that it was harboring a nest of white, cottony lumps tucked under its leaves and around its leaf axes. Or maybe there was no visible evidence of infestation—only tiny white eggs which later hatched into a colony of sluggish little puffs. These are mealybugs and, along with spider mites, are the most prevalent of all the pests which plague indoor gardeners. Although they're very fond of coleus, African violets, cacti, and ferns, mealybugs aren't selective diners; they'll happily feed on almost any plant, devitalizing and eventually killing it if vigorous efforts aren't taken to eradicate them. There are three types of mealybugs that commonly afflict houseplants—the long-tailed, the short-tailed and the root mealybug. The last one favors cacti and feeds on the roots of potted specimens. Reproduction is usually by egg-laying, although some types give birth to living young. Eggs are carried by the females in cottonpuff-like sacs. The typical female would put a hen to shame, for she is capable of laying up to 500 eggs at a time. Our cozy interior environments are ideal nurseries for these eggs: the hatching cycle, which is normally two to three weeks outdoors, is accelerated to about eight days indoors. New broods, which are oval, smooth, six-legged and yellowish, emerge with voracious appetites and begin to extract sap immediately. A few days later, nourished and energized by pilfered sap, they start secreting the white, waxlike mealy coating which is characteristic of the species. Males produce a white "cocoon" which they wrap around themselves and metamorphose into winged insects capable of moving from plant to plant seeking a good woman to settle down with and raise a new brood.

Damage.

The continuous tapping of a plant's sap reserves first causes the foliage to become pale, then yellow and, finally, brown. New growth emerges stunted, twisted and lacking vitality. Mealybug excrement, called honeydew, may attract ants which feed on it and it also provides a growing medium for sooty black mold.

Favorite Targets:

African violet, cacti, citrus, coleus and ferns.

Controls.

Anyone who has ever read an article on houseplant care knows that accessible mealybugs may be safely removed from

Individual mealybugs can be safely removed with a cotton swab dipped in alcohol.

plants with a cotton swab dipped lightly in alcohol. The advice is sound and effective, as far as it goes. Alcohol cuts right through the mealybug's protective wax coating like acid and shrivels the little critter up like a dried prune. But the swab-and-alcohol treatment is only a temporary measure because the microscopic mealybug eggs, which have probably been laid by the hundreds, are missed. Stronger measures are called for to do a thorough job of annihilating future colonies. One of the safest and most effective of these is a dip comprised of three tablespoons of Malathion liquid concentrate and two tablespoons of household detergent (to serve as a spreader) dissolved in a gallon of lukewarm water. Immerse the plant down to the soil line and swish the plant around a few times. This procedure may have to be repeated in ten days to two weeks. Malathion should *not* be used on succulents, cacti and ferns, which can be seriously damaged by this particular pesticide. For these plants, and to eradicate soil mealybugs which feed on roots, use a systemic which is applied to the soil. Isotox is one of the best of these.

NEMATODES.   Also called eelworms or wireworms.

Description/Habits.

Nematodes are microscopic animals resembling worms. They exist naturally in the soil. The three types which concern gardeners are leaf nematodes (*Aphelenchoides fragariae*), root-knot nematodes (*Heterodera marioni*) and stem/bulb nematodes (*Ditylenchus dipsaci*). Nematodes enter the cell network of plants by feeder roots or through the pores (stomata) of foliage. Females lay some four hundred eggs at a time and these require an incubation period of only a few days.

Damage.

Leaf nematodes cause severe, splotchy browning of foliage in many plants and flowers and brownish-black bands of discoloration on fern fronds. Misshapen new leaves and irregular lumps on stems and foliage are also signs of nematode damage. Afflicted plants are often distorted and, in the final stages, leaves begin to shrivel and drop as damage moves up through the plant. Root-knot nematodes invade the tender tips of feeder roots, and eventually the entire root mass is transformed into a deformed clump of galls and knots. Feeding is accomplished by puncturing the plant tissue with a hollow,

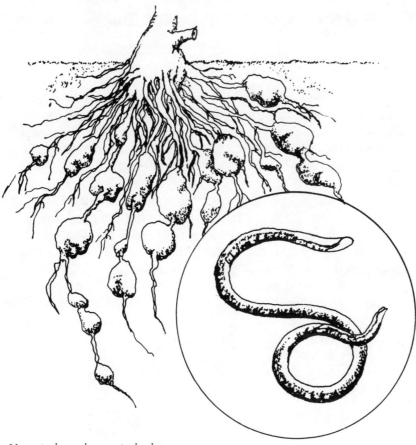

Nematode and nematode damage

syringelike organ and sucking up plant sap, which ultimately devitalizes and kills the host plant.

Favorite Targets:

African violet, citrus, crassula, some palms.

Controls.

Prevention of nematode infestation should be your first step and this can be achieved by potting up plants in pasteurized packaged soil and rooting cuttings in sterilized sand. Nematodes suspected of lurking in garden soil used indoors in containers may be eliminated by saturating the soil with a nematocide such as V-C-13. Before modern chemistry developed nematocides to combat nematodes, gardeners mixed up a solution of one teaspoon of chlorine bleach to one quart of water and immersed nematode-infested plants in this solution up to the soil line for an hour.

RED SPIDER MITES. Also called "Two-spotted mites."

Description/Habits.

These persistent, destructive and, unfortunately, commonplace pests barely qualify for their "red spider" cognomen by their inept efforts to spin webs. The results of their web-making would put any *bona fide*, red-blooded spider to shame and this is one of the techniques you can use to determine "if you got 'em or if you don't got 'em." Is the web feeble and scraggly with tiny white specks (empty mite skins) on some of the strands? Then you got 'em. Is the web symmetrical and finished, like most spider webs you've seen all your life. Then you don't got 'em. What you have is a common spider who liked the way you were forgetting to mist your plants, thus providing a warm, dry place to build a trap for unsuspecting insects. Another good method for detecting mites is to hold a sheet of white paper under a leaf and tap the leaf several times. If the plant is infested, some of the colony ensconced under the leaves, away from direct light, will be dislodged and fall on the paper. Any tiny red dots scampering around on the paper are red spider mites and you've got your work cut out for you.

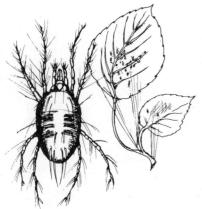

Red Spider Mite

Damage.

Of all the indoor plant parasites, red spider mites are among the most dangerous because of their virtual invisibility until their colonies are well established and the host plant is, consequently, seriously imperiled. Both newly hatched and mature mites pierce the epidermis of foliage and suck up sap in an almost continuous orgy of feeding. The leaves first exhibit evidence of this vampirism by paling out. Then tiny red and yellow spots appear where the leaves have been pierced and the surfaces of the leaves look dull, as if covered with a layer of fine powder. This is a combination of webbing, eggs and mite carcasses. Leaves, sometimes still green, begin to fall until the entire plant has been devitalized and defoliated.

Favorite Targets:

Aspidistra, cissus, citrus, dracaena (particularly, *D. marginata*), schefflera and podocarpus. Actually, few plants are immune from spider mite attack.

Controls.

Of all plant pests, probably red spider mites move from plant to plant with the greatest ease, ultimately infesting an entire collection. Mites can be carried on a breeze blowing through an open window or by a plant owner touching an infested plant and inadvertently transporting the pest to another host. Or the mites simply move on their own from plant to plant in a collection that is crowded together. Obviously, then, if you have a plant that has a confirmed case of mite infestation, immediate isolation is crucial.

Red spider mites consider a hot, dry environment an open invitation to colonize. They abhor moist, humid conditions, so frequent misting or syringing may discourage them from invading your collection or, at least, keep them in check. One of the first steps to take in eradicating a mite colony is simply to submerge the foliage under a stream of lukewarm water from the tap. Larger plants and trees can be hosed off outdoors. For a slightly stronger treatment, put one tablespoon of Ivory soap bar shavings in a pint of hot (not scalding) water, shake until the shavings are dissolved, pour the soapy mix into a hand mister and cover the foliage, paying particular attention to the undersides. Leave the suds on the foliage until almost dry, then rinse completely with cool water. Or mix up the soap as a dip in the sink or a bucket and swish the foliage around in it several times. If this proves effective, bully! Anytime you don't have to resort to pesticides, you're ahead of the game. Unfortunately, spider mites have an uncanny instinct for survival and if the soap-and-water method fails you may have to use a chemical compound that doesn't pussyfoot around. First, try the most ecologically safe of the pesticides, Malathion, as a dip or spray. Combine three tablespoons of household detergent in a gallon of lukewarm water and add three tablespoons of this evil-smelling concentrate. Either spray or dip the plant. Remember that Malathion is potentially harmful to ferns and cacti. For these and other sensitive species use a systemic in granular form dissolved in the pot soil. Other effective pesticides include Tedion, nicotine sulfate, Difocal and Tetradifon.

Scales

SCALES

Description/Habits.

Scales come in many shapes and sizes (see illustration) and are divided into two groups: *armored*, or hard-shell; and *tortoise*, or soft-shell. Scales live, feed and give birth under their protective shells, or scales, safe from other insect predators and, usually, from insecticide sprays. Armored scales manufacture their shells from thin wax threads and cast-off skins after molting. Their scales, which aren't attached to their bodies, serve as impregnable fortresses, sealing them in against danger. Tortoise scales differ from armored scales only in that their scales develop naturally and are not separable. Scales proliferate rapidly into destructive colonies. New broods may be either hatched from eggs or born live. They move out from under their mothers' shells soon after birth to stake out their own feeding grounds. Unlike female scales, which remain virtually anchored to one spot, males move around to mate and, after they have molted twice, metamorphose into saffron-colored, two-winged insects.

Damage.

Scales rob plants and flowers of vitality by extracting sap from cells. Their honeydew excretions attract ants and flies which feed on the tacky substance. Honeydew also provides a growing medium for sooty black mold.

Favorite Targets:

Aralia, bromeliads, cacti, citrus, crassula, ficus (especially *F. benjamina*), schefflera and podocarpus.

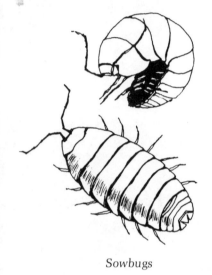

Sowbugs

Controls.

Scales can be removed using the dull edge of a knife or scraped off with a thumbnail. The spot where they were removed should be scrubbed with Ivory soap and water to kill off any eggs or young that may be present. Serious infestations are best treated by using a systemic such as Isotox in the soil. Malathion sprayed on the foliage kills exposed young who haven't had time to form protective shells. Systemic granules in the soil work best for sensitive ferns and succulents, which are both frequently damaged by Malathion.

SOWBUGS.   Also called pillbugs and roly polys.

Description/Habits.

Sowbugs call to mind armadillos with their grayish, segmented bodies and their protective reflex action of rolling into a ball when disturbed. Most have seven pairs of legs and are seldom larger than a quarter of an inch long. Females lay about twenty-five eggs at a time. These are carried in a ventral pouch called a *marsupian*, which is home for the young sowbugs for some time after they hatch. Compost piles and heaps of garden debris where it is always damp are favorite hiding places for sowbugs. They are brought indoors in canned stock which has been sitting in the lath house at the nursery or in houseplant pots which have been set outside on the ground to air.

Damage.

Sowbug destructiveness is vastly overrated. They are scavengers who feed primarily on decaying vegetation and garden debris. They can, however, damage young seedlings and plant roots.

Favorite Targets:

Succulent young seedlings of herbaceous plants.

Springtail

Controls.

The simplest method of eradicating sowbugs is to drown them out by submerging an infested pot in a bucket of water. Those flushed to the surface can be scooped out with a spoon and discarded, A drench of Malathion will get the job done in a hurry but really isn't necessary.

SPRINGTAILS

Description/Habits.

Like the sudden visitation of fungus gnats, the appearance of springtails is a sure sign that you've been overactive with the watering can. Under magnification, springtails resemble very homely fleas. They are usually dark and hardly more than 1/20th of an inch long. They are not genuine insects but are technically "proto" or "pre" insects. You'll first notice them leaping around on the topsoil of a plant you're watering. These acrobatics are accomplished with a forked tail which acts as a catapult, propelling them through the air.

Damage.

Any leaves near the soil line are endangered by springtails, which chew small holes in tender foliage.

Favorite Targets:

All moisture-loving plants.

Controls.

A Malathion drench eliminates springtails in short order.

THRIPS

Description/Habits.

Thrips

With the growing popularity of plant collecting and the cultivation of newer species over the last few years pests have appeared that few, if any, indoor gardeners even knew existed. Thrips (the name is both singular and plural) is one of these. It is a tiny, mean, voracious predator belonging to the *Thysanoptera* order, which means "bristle-winged" (see illustration). Like whiteflies, thrips are active and mobile, effortlessly flying from one plant to another, which makes control by spraying almost impossible. Adults are thin, brownish or black and sometimes accented with grayish markings. Nymphs are white or yellow. Sharing another similarity with whiteflies, they scamper across leaf surfaces or fly or leap away when disturbed. Nymphs molt four times and their coloration changes from yellow to amber to brownish-black at maturity, and they develop four bristled wings.

Damage.

Thrips are sap-suckers who first slice a ragged incision in the surface of a leaf, then insert their hypodermic tongue to feed. Females open a wound in a leaf and lay their eggs in the opening. Ten days or so later, tiny white nymphs with insatiable appetites begin to eat their way through the foliage. Thrips damage is easily diagnosed. Where thrips activity is present, leaves take on a silver-streaked appearance. Large areas of the leaf surface develop pale, papery scars where the sap has been extracted and noticeable blisters appear where the females have deposited their eggs. Finally, trails of black or brown specks of excrement are visible. In some cases, leaf tips curl and wither and, in the case of flowering specimens, buds often drop or open to reveal a misshapen, discolored bloom.

Favorite Targets:

Aralia, citrus, *Ficus nitida* (Indian laurel), most flowering plants.

Controls.

In the author's experience, 100% control of thrips has been achieved only through the use of a systemic pesticide in granular form, scratched into and then dissolved in the soil by irrigation. This gets the nymphs feeding inside leaves as well as the adults. But with thrips a two-pronged attack is recommended—something for a quick knockdown of thrips while you're waiting for the systemic to be absorbed and translocated throughout the plant's system. This can be a Malathion dip or spray as described for use against red spider mites, or Cedoflora.

WHITEFLIES.   Also called "Greenhouse whitefly."

Description/Habits.

These small (1/20″), mothlike parasites rank with spider mites in their survival capabilities under a barrage of chemical attacks and verbal abuse from plant lovers unlucky enough to have a plant infested with them. Adults achieve their white, mealy appearance by secreting wax which serves as a protective coating. Like most pests sent to plague us, the females are prolific breeders and, worse luck, the cycle for maturation of each generation is a mere five weeks. Eggs are attached to the undersides of leaves where the hatched nymphs, which are infinitely small, pale-green and scalelike, begin to feed hungrily. As the nymphs mature, they change color from pale, translucent green, to white, then to yellow, when the skin opens to release the adult. Upon achieving adulthood, they "earn" their wings and join the other adults to rise, *en masse*, to hover over an infested plant when alarmed and settle back on their victim once the disturbance has passed. Inevitably, some move to nearby plants to establish new colonies. Their honeydew excretions attract flies, ants and encourage the growth of sooty black mold.

Damage.

Whiteflies, like other sucking pests, remove sap from plants. Eventually the plant no longer has sufficient nourishment to survive, let alone thrive. As the vital juices are drained away, foliage begins to yellow and eventually drops away. Soon the plant begins to wilt and death is unavoidable unless help is provided.

Favorite Targets:

Azalea, coleus, citrus, fuchsia and lantana.

Controls.

Because of the pest's mobility, sprays and dips are seldom 100% (or even 90%) effective. Inevitably, a few of the airborne little suckers will evade the lethal mist. That is why control is so difficult to achieve and why many horticulturists and botanists recommend discarding whitefly-infested specimens. Sprays that are effective against the pest are Malathion or Rotenone. For ferns, use nicotine sulfate, ¾ teaspoon in a quart of warm water to which you've added a tablespoon of household detergent as a wetter-spreader, or use systemic granules dissolved in the soil. Probably the most successful technique is to place the afflicted plant in a garbage can, suspend a Vapona strip from the edge or lid with a wire, close the lid tightly, and leave the plant overnight.

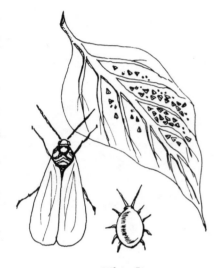

Whiteflies

Diagnosing
Parasite Damage

Condition	Probable Cause	Remedies
Leaves stunted and distorted, tips discolored. Buds deformed, flowers pale, unnatural in color.	Aphids	Soap and water may eliminate aphids. If they return, use Malathion-50 on all but ferns, cacti and succulents. Nicotine sulfate is safe for these.
Leaves and stems chewed, scarred and new tender leaves missing.	Cockroaches, Crickets or Earwigs	Malathion-50 handles these chewing insects, but use nicotine sulfate on ferns, cacti and succulents.
Foliage distorted with scabrous, lumpy surfaces. Buds black and deformed.	Broad or Cyclamen mites	Use a systemic in granular form dissolved in the soil.
White dots on undersides of leaves visible on upper epidermis.	Leafhoppers	Use systemic pesticide in granular form in soil. Malathion-50 spray on all but ferns, cacti and succulents. Use nicotine sulfate on these.
Leaves appear to have linear and/or serpentine mines, blotches with dead tissue at center.	Leaf Miners	Remove badly mined leaves, which may still contain pests. Use a systemic in the soil.
Leaves covered with tiny puffs of cotton. Tacky substance on leaves and stem. Foliage turning yellow.	Mealybugs	Pick off adults with alcohol-dipped swab. Spray to destroy eggs and young with Malathion-50, except on ferns, cacti and succulents. Use nicotine sulfate on these.

(continued)

Condition	Probable Cause	Remedies
Foliage develops brown splotches. New leaves are misshapen. Irregular lumps appear on stems. Fern fronds show brownish-black bands of discoloration. Plant vitalized.	Nematodes	Use a nematocide, such as V-C-13.
Wispy, incomplete webs appear between leaves or leaf tiers. Red and yellow flecks develop on foliage. Leaves lose their sheen, vitality and rich color.	Red Spider Mites	Malathion-50 spray for all but ferns, succulents and cacti. Use a systemic in the soil for these, or spray with nicotine sulfate.
Leaves and stems covered with small brown, immobile lumps and sticky, clear substance. Foliage lacks natural bright green color. Plant seems drained of vitality.	Scales	Remove by hand. Use a systemic pesticide in granular form dissolved in pot soil.
Leaves have a silver-streaked appearance. Surfaces show pale, papery scars. Trails of black specks visible. Buds show color but fail to open. Petals edged with brown.	Thrips	Systemic pesticide used in spray and granular form.
Clouds of white insects hover around plant. Foliage turning pale or yellow. Plant looks on verge of general collapse.	Whiteflies	Best treatment is systemic pesticide dissolved in pot soil.

Insect Predators
and their Targets

Aphids

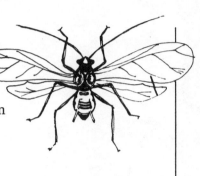

Begonia
Calendula
Chrysanthemum
Delphinium
Fern
Ivy
Pittosporum

Mealybugs

African violet
Amaryllis
Begonia
Chrysanthemum
Citrus
Coleus
Crassula
Croton
Dieffenbachia
Dracaena
Ivy
Fern
Gardenia
Kalanchoe
Palm
Poinsettia

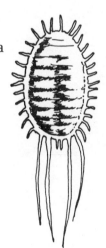

Broad/Cyclamen Mites

African violet
Cyclamen
Delphinium
Geranium

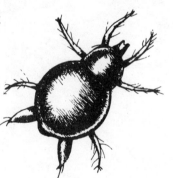

Nematodes

African violet
Begonia
Chrysanthemum
Crassula
Gesneriads
Palms

Leaf Miners

African violet
Azalea
Ficus
Palm
Privet

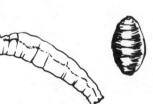

(continued)

Red Spider Mites

Ageratum
Amaryllis
Aspidistra
Begonia
Cactus
Citrus
Dracaena
Ivy
Schefflera

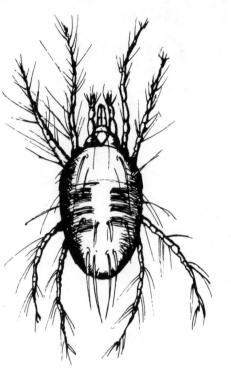

Thrips

Aralia
Avocado
Begonia
Chrysanthemum
Citrus
Croton
Cyclamen
Fern
Ficus nitida
Gloxinia
Jerusalem cherry
Orchid
Palm
Privet

Scales

Aloe
Aralia
Avocado
Bromeliad
Cactus
Citrus
Croton
Dracaena
Fern
Ficus
Gardenia
Ivy
Palm
Privet
Schefflera

Whiteflies

Ageratum
Avocado
Azalea
Begonia
Calendula
Coleus
Fuchsia
Lantana
Privet

Chapter 5

surgery and intensive care

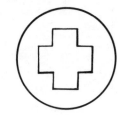

I f you plan on building a varied collection of plants, or have a dozen or so now, one or more may one day need an emergency operation. Surgery is usually one step in an overall program of plant medication and care. It is seldom a cure-all, but it often stops the progression of a disease by removing the source of decay.

OCCASIONS FOR SURGERY

Root Rot, a common fungus-oriented disease brought on by prolonged overly wet soil conditions, can be curbed by surgery if caught in time. The plant is knocked out of its pot and all of the soil is gently removed. Holding the roots under a stream of lukewarm water helps here. Be sure to support the root ball in one hand to prevent the weight of the soil from breaking it off at the crown. Let the water and soil trickle through your fingers.

Once all soil is removed, examine the roots. Black or brown mushy portions are dead and must be removed. Healthy roots are firm and white. Lay the patient on a paper towel and, using a sterile (or new) single-edge razor blade, mat knife or other sharp blade, cut away diseased portions of roots well above the decayed tissue. Scissors are not recommended since they tend to crush and pinch as they cut.

Immediately discard the rotted tissue, rinse the roots again, pat dry and dust with a fungicide. Then dip the trimmed root ends in a rooting hormone and pot up with *new* soil and a *new* pot.

Your chances for success in saving a plant with root rot are excellent if only 10% of the roots are diseased, good if 25% are affected, and fifty-fifty if half the root system has deteriorated.

Place the recuperating patient in a location that gets medium natural light (no direct sun!) and has a more or less

Root-rot Surgery/ Therapy

Remove soil from roots under tap.

Trim off decayed roots.

Dust roots with fungicidal powder.

Dip roots in rooting hormone.

Pot up in new soil, new container.

constant temperature. Withhold water and food until new growth appears. If the plant is going to pull through, you should see signs in about ten days to two weeks. From this point on, be stingier with water than had been your habit before the need for surgery.

Stem/Leaf Rot, if not too severe, can sometimes be halted by slicing out decayed areas, again with a clean, sharp instrument, and dusting the area with a fungicidal powder. These cuts should callus in a day or two, and if you see no further evidence of decay after a week or so, you're a brilliant surgeon. Remember, the development of the fungus was the result of inadequate care—too much water, stagnant, humid air, allowing the leaves to go to bed at night wet, etc. Determine what you were doing wrong and attack the problem from that position.

Leaf, Branch Removal, Trimming.

Removal of single leaves is not really a surgical procedure, but it should be done with care and precision. Small leaves can simply be pinched off by grasping the petiole between the thumb and finger and pinching. Larger leaves and branches must be smoothly cut away with a sharp instrument to avoid damaging or breaking the stem in the process.

When Should Leaves/Branches Be Removed?

This is partially a subjective question. Only you can decide whether you should remove leaves that are still healthy although they have been damaged by careless handling or have a hole chewed through them or have sunburn spots in the center. Personally, the author usually opts for removal, if the structure and/or appearance of the plant won't suffer. It seems more logical to have the plant shift its energy and food from sustaining an inferior leaf into producing and supporting a fresh, new, undamaged one.

There are more practical reasons for removal of leaves, however. A common practice among those who know the growth habits of particular species is to pinch out new growth on specimens to force them to fill in better down below and increase the size of older leaves. Most plants—philodendrons, for example—are more concerned with new growth and expend most of their energy in gaining height instead of girth. That is why you see so many split-leaf philodendrons that are

Slice out decayed area and dust with fungicidal powder.

Dead branches should be removed to enhance appearance of plant.

nearly all vine, topped by a few insignificant leaves. If these plants had been pinched back every couple of months the first year, they would have retained most of the foliage further down the vine and even popped out new leaves in between. When a plant is frustrated in its attempt to grow up it will grow sideways, going back and filling in the areas on branch and stem and trunk it had no time for before.

Not all plants respond to this technique. Some which do are dracaenas, ivies, philodendrons, cordylines, and scheffleras.

Any leaf that is harboring insects between the upper and lower epidermis, such as leaf miners, or is afflicted with fungus disease, or is so badly encrusted with scales or mealybugs that it's already moribund, should also be removed.

Tree branches that have lost their foliage and are obviously dead can either be left or removed. Usually, the appearance of the tree is improved if they are removed. Once dead, of course, they will never again produce foliage. Also, branches that have been partially torn from the trunk by accidents or careless handling should be cut away so that the exposed tissues are not invaded by disease. Dust the wound with fungicidal powder to protect it until a callus forms.

Browned-off or discolored tips and margins of foliage can safely be given "cosmetic surgery" with scissors. Some horticultural experts denounce this practice as needlessly traumatic. They are of the opinion that deteriorating foliage should be allowed to wither and drop on its own. In the author's experience none of the hundreds of plants spruced up in this

Pinching out new growth often encourages bushier growth.

This Crassula (jade plant), badly infested with scales, was cut back to the naked trunk three weeks prior to this photograph.

Foliage on this Ficus roxburghii was severely damaged by insects and sun. Two weeks after cutting back, new foliage emerged from dormant buds.

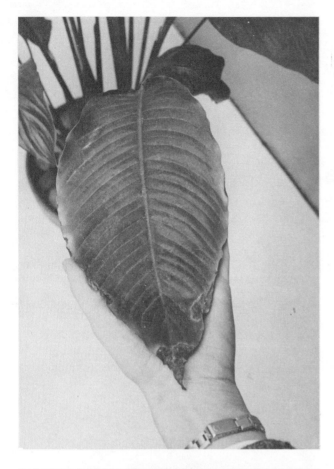

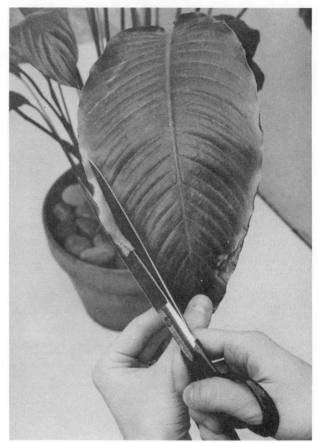

Browned-off leaf tips and margins detract from a plant's appearance.

The dead tissue can safely be removed with a pair of shears.

Appearance of leaf is now considerably enhanced.

way have suffered any from the operation. It is good preventive medicine to remove necrotic tissue before it provides a haven and breeding ground for disease. Besides, a plant's appearance is important and it always looks better after a beauty treatment.

Drastic Measures.

Occasionally, a plant may be so heavily infested with predators that seemingly no amount of doctoring can completely eradicate the pests. You think you have conquered the critters one week and the next week they're back. After months of this, the foliage may be badly ravaged and the plant is more of an embarrassment than a joy to you.

In many cases, there is a radical step you can take to save and rejuvenate the plant if you can wait until spring or summer. Cut the plant back and let it start again. Plants with foliage right down to the soil line should be cut here. Other plants and trees should be cut back just under the foliage.

Plants which respond well to this surgery and recover in one growing season are: all citrus, *Cordyline*, *Dizygotheca elegantissima* (false aralia), *Dracaenas*, *Fatshedera* (tree ivy), *Ficus decora* (rubber tree), *Monstera deliciosa* (split-leaf philodendron), *Pittosporum tobira* (mock orange) and *Pseudopanax*. Most of the above, after having been cut back, come back strong *and* branch attractively as a bonus.

Those plants which shouldn't be cut back are: cacti (although most other succulents can be), *Araucaria heterophylla* (Norfolk Island pine, or star pine), *Howeia belmoreana* (Kentia palm) and *Beaucarnea recurvata* (bottle palm, or elephant-foot tree). Bottle palm is a special case. It is an extremely slow grower, but lopping off the top tuft of leaves will often promote attractive clumping and branching. The problem is that it usually requires years to achieve this, and you're stuck with a topless plant all this time.

New Plants from Old.

Many plants that have deteriorated from attacks by insect predators, or have had their stems snapped in two by accident, or have just become so leggy that they've lost most of their appeal, can be given a second chance by cloning or, with woody-stemmed plants, by air layering. Cloning is asexual or vegetative propagation, since you use a part of the "vegetation" instead of a seed.

Drastic Surgery and Treatment

Severely-damaged plant is cut back and moistened for removal from container.

Plant is removed from old container.

Some roots are pruned to stimulate new root development.

Root ball is dipped in rooting hormone solution.

Plant is potted up in new container.

Soil is pressed gently but firmly around roots.

Pot is banged against solid surface to settle soil around roots.

Newly-potted plant is watered, then set aside to recover.

Stem Cloning.

Softwood stems (green stems from plants like dieffenbachia) are the easiest to root. Cut the stem in several pieces, each time just below a node, or leaf joint, which is where roots will form. Insert the cutting, node down, in a rooting medium you've mixed of either half sand and half shredded peat moss or half milled sphagnum moss and half vermiculite. The container can be anything from a coffee can to a bulb pan, but should have at least one drainage hole so excess water will drain out.

Cuttings need warmth and humidity, so after you've sprinkled a little water over them put the container in a plastic bag and tie it shut. Poke a couple of holes in the top so that some of the moisture can escape to avoid the formation of mold, and set the cuttings in a warm spot, but never in direct sun which would cook the bejeebers out of them.

Check the rooting medium every two or three days for moisture content. It should be just damp, never wet. You may not have to water the cuttings for some time because the moisture which forms on the inside of the bag will drip down onto the cuttings from time to time in a continuous recycling effect.

With softwood cuttings rooting time can be anywhere from ten days to a month. Hardwood cuttings may require up to three months to produce satisfactory, viable roots. When roots are an inch long, remove the cuttings and pot them up.

Leaf Cloning.

Succulent leaf cuttings from such plants as crassula, peperomia and sansevieria, can be induced to root in the same manner as stem cuttings. Sansevieria leaves should be cut into short lengths and inserted into the rooting medium to a depth of about a third of their length. Leaves with petioles, or stems, should be buried—stem down—about the same depth. Crassula leaves (and stems) have a tendency to rot if buried too deeply. Simply pressing them into the topsoil is usually sufficient to promote rooting. Most leaves root faster if dipped lightly in a rooting hormone before planting.

Air Layering.

A hardwood-stemmed plant or tree which has gone leggy, or one which you'd like to have two of, can be air layered. This technique involves cutting a wedge from the trunk or stem

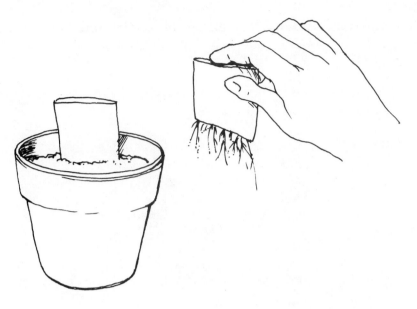

Bury cuttings a third of their length. In a short time they will sprout roots.

where you want to produce roots. The incision should go deep enough to reach the core of the stem. The wound is packed with damp sphagnum moss which should be tightly secured in place with cord or gardening twist-ties. The idea is to keep the cut from forming a callus (tissue which grows over a wound to seal it), which is the plant's way of protecting itself from loss of sap. As long as moist moss is kept in contact with the wound, it will not callus. The moss is then covered with plastic foodwrap which is taped or tied to create an airtight seal top and bottom. This seal helps keep the moss from drying out.

It may take as long as three months for the development of sufficient roots to permit separation of the air-layered portion from the "donor" plant. During this waiting period, check the moss weekly to make certain it's still damp. When roots have nearly filled the moss (you can see this through the wrap), you can separate the air layered section and pot it up.

If the plant being air layered has heavy top growth, support the stem by tying it to stakes pushed into the soil. Once the stem is cut for the air layering operation, it may not be strong enough to support the weight of heavy top foliage without bracing.

You should also stake newly planted air-layered sections for a month or two, to give their roots time to develop enough so that they can support the plant. Otherwise, they may topple over and uproot themselves.

Keep the newly planted section a bit on the dry side for a few weeks. This will encourage faster development of roots to go looking for moisture.

The plant from which the air-layered section was removed will usually "stump sprout" where the cut was made, giving you two plants for the price of one.

Sanitation and Disinfection.

The spread of parasites and plant diseases from sick to healthy plants should be a constant concern. Check your plants thoroughly at least once a week for signs of predators or incipient disease. When you find something amiss, take immediate action. Plants have no built-in defenses. They can't cure themselves.

After isolating and treating a diseased or infested specimen, scrub your hands with soap and water. The danger of transmitting infection or infestation to healthy plants is easily avoided by this simple precaution, and those that follow.

All cutting instruments and other tools used in treating sick plants should be disinfected immediately after you've finished. Plastic pots and instruments can be cleaned with hot soapy water and a little Clorox. Scissors, knives and razor blades used to treat the diseased plants should be dipped in copper napthenate to disinfect.

Finally, clean up and discard debris from surgical operations as soon as you've finished. Don't add these trimmings to the compost pile since the infestation they are carrying would rapidly multiply.

Chapter 6

pesticide, insecticide and chemical controls

It would be wonderful if we never had to use chemicals for pest and disease control—if we could just allow the natural enemies of destructive insects to do the job which nature intended. This simply isn't practical indoors. If an invasion of parasites is ravaging a prized plant which you've raised lovingly through the years, and you've exhausted all the organic methods of control without success, it becomes a matter of choice: you either resort to something stronger or you kiss the plant goodbye—and maybe a great many of your other specimens as the infestation spreads.

Pesticides and insecticides have been given a black eye by ecologists and organic gardeners. Anyone who advocates their use is all but drummed out of the human race by persons of this persuasion. There is no question about it—chemical pesticides have created a serious imbalance in the ecosystem, killing billions of beneficial insects along with the destructive ones. Pesticides have poisoned the air and polluted rivers and streams to catastrophic levels. The use of DDT, the worst of all insecticides, has cost us dearly and we'll be paying that debt to Mother Nature for the next fifty years, even though we've all but stopped using DDT.

But there are a few pesticides which are relatively safe to use, if handled intelligently: chemicals which are degradable (which means they break down in the soil and their "kill potential" decreases to nil after the process is complete); substances which kill only the pests they were designed to kill and pose no threat to man or animal. The key phrase where chemical pesticides are concerned is "if handled intelligently."

Even though it is often necessary to resort to chemical warfare to control pests and/or disease, you should make a conscious effort to select the least toxic preparation that will get the job done. It is purely and simply irresponsible and potentially dangerous to use the overkill approach in dealing with a few pests. The following chemical pesticides/insecticides should be avoided altogether. They are chlorinated hydrocarbons which pose real threats to the ecological balance of our planet. You don't need indiscriminate killers to control pests on your plants. There are plenty of safer substitutes that are equally effective in eradicating plant parasites.

Persistent Chlorinated Hydrocarbons

Aldrin (Banned by the EPA)
BHC (benzene hexachloride and lindane)
Chlordane (Banned by the EPA)
DDD (TDE)
DDT (Banned by the EPA)
Dieldrin (Banned by the EPA)
Endrin
Heptachlor (Banned by the EPA)
Methoxychlor
Toxaphene

No pesticide/insecticide is totally safe or innocuous. Every time you use one there is some risk of inhaling irritating or noxious fumes, or killing beneficial insects with drifting mist. But a number of chemicals, some of them organic, are relatively low in toxicity and danger to man, animal and the balance of nature. Some are stronger than others, but none of them is as catastrophically dangerous as the "doomsday machine" chemicals labeled "persistent chlorinated hydrocarbons."

Malathion is a contact insecticide which deals successfully with nearly all sucking insects which infest houseplants —aphids, leafhoppers, mealybugs, red spider mites, some scales, thrips and whiteflies. It can be used as a dip, drench or spray on all plants except ferns, cacti and certain other succulents whose tissues are sometimes damaged by it. Check the label for specific plants sensitive to Malathion. For these species, a good substitute is a systemic pesticide in granular form dissolved in the soil.

Cedoflora, a pleasant-smelling concentrate of hemlock and cedar oils, petroleum and soap, is effective against broad and cyclamen mites, mealybugs and scales.

Isotox. Over the last ten years, during which the author has grown hundreds of plants and trees indoors, the safest, surest and most trouble-free of all the chemical preparations developed for the control of sucking insects (which covers 90% of all houseplant pests) was determined to be a systemic poison in granular form, such as Isotox.

The sprays, such as Malathion, and even some systemic sprays, give you a quick knockdown of adult predators on a plant, but these alone seldom do the job. Pest eggs are not always killed along with the adults, so you have to do this unpleasant job two or three times over a period of weeks to eradicate newly hatched pests. Also, some predators often are able to crawl deep inside buds and leaf folds and escape sprays. Whiteflies and leafhoppers are agile enough to flee from danger when you approach with the applicator. Finally, sprays are ineffective on such pests as leaf miners, which feed inside the leaf or adult scales, the shells of which usually protect them from sprays.

A granular systemic is applied by scratching a circular trench an inch deep in the topsoil with a label marker or screwdriver all the way around the stem of a plant. Following label instructions, the granules are poured into the trench, covered with soil, then dissolved to start the treatment by irrigation (which should be done carefully to avoid washing up the granules, which give off an unpleasant odor when wet).

Water containing the solution filters down to the feeder roots. They take it up and disperse it throughout the plant's system. This is why such pesticides are called "systemics." From that point on, for many months, any pest that attempts to feed on the plant's sap will be killed.

There are some dangers with the use of a systemic pesticide that should not be taken lightly. Both the soil and the entire plant contain enough poison to cause severe nausea and even death in children and animals, both of which are sometimes fond of eating dirt or sampling foliage. If you have either small children or gnawing puppies, you may want to forget about systemics and stick with less dangerous preparations.

Nicotine sulfate, widely sold for years as "Black Leaf 40," a 40% solution of tobacco extract, is the alternative choice insecticide for use on plants which may be damaged by the use of Malathion and other chemicals. It is a contact insecticide that is very effective in controlling sucking insects.

Nine times out of ten, you can easily combat *any* of the sucking and chewing insects with just these four insecticides, and do it with little danger to yourself, your family, your pets and the ecology, assuming you follow label directions and observe common-sense safety precautions.

There is one other type of product on the market which is recommended largely because of its convenience, not because of its effectiveness or relative safety, although most contain very little of the active ingredient (insecticide). This is the "bug bomb."

Aerosol sprays (bug bombs) are effective in controlling a broad range of pests which attack houseplants. The insecticide is dissolved in a liquified gas (methyl chloride or Freon) and kept under pressure. When the valve is depressed, the solvent vaporizes, leaving the insecticide suspended in the air. There are two dangers involved in the use of bug bombs, one to the plant and one to you. If the valve is held too close to the plant, the propellant gas can freeze the foliage. Inhaling the insecticide, it should be obvious, is not the healthiest of things you can do for your body. Bombs should never be used indoors where the spray can drift to dishes, foodstuffs, pets, infants, etc., and they should never be used in an enclosed, unventilated area, where you'll get as much of the mist in your lungs as the plant gets on it foliage.

The best way to use aerosol sprays is to put the plant to be treated in the bottom of a plastic bag, spray, and quickly seal the top. A garbage can with a lid can also be used. For maximum results leave the plant sealed inside for at least fifteen minutes.

If you spray a large plant or tree outdoors, wear a respirator ($1 to $2 at paint stores) and spray only on calm days to prevent the insecticide from drifting away or blowing back in your eyes.

Pesticide Footnotes—or Questions in Search of Answers.

In their anxiety to avoid possible lawsuits, pesticide manufacturers often go to ridiculous extremes with their label warnings: "Do not inhale vapor of this product!" "Keep out of reach of children!" "Do not eat or smoke while using this product!" We certainly aren't going to empty the contents

For application of system granules, scratch inch-deep trench in soil.

Following label instructions, pour in granules.

Cover with soil, firm and water to dissolve granules.

of a bug bomb in our nostrils, or give a can of bug spray to our kids to play with, or even consume a peanut butter sandwich while killing off the aphids on our schefflera.

We all know that deadly chemicals are not for human consumption. What the chemical companies *should* be telling us is not the obvious, but information such as that *no-pest strips* contain Vapona (DDVP), a dangerous chlorinated hydrocarbon. The chemical is imbedded in wax and is released into the air as the wax evaporates. They should be telling us that if you

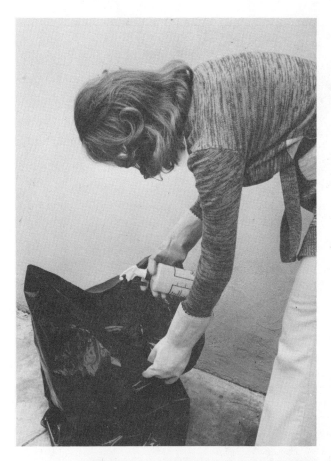

Plants can be safely, effectively sprayed by setting plant in plastic bag.

Apply spray with bag pulled up around plant.

Secure top of bag with twist tie.

Use a respirator when spraying noxious pesticides.

hang a no-pest strip inside your home, your family and food-stuffs, along with the bugs, are exposed to the vapor.

They should also be telling us that many *shelf-lining papers* that kill bugs do so because they are impregnated with chlordane (banned by the EPA) and lindane, two more chlorinated hydrocarbons. Do you want your dishes and foodstuffs bathed in waves of chlordane and lindane?

And they should be telling us what kind of logic prompts them to say we can use their *insect foggers* in our backyards and go right out and have a picnic, yet the killing effect of the spray on insects (bees and ladybugs, too) goes on for hours. Do the chemists who make the foggers possible use them at *their* family picnics, knowing what makes the vapor "kill bugs dead"? It's doubtful.

The only words of caution the author has are: when the manufacturers of insecticides come right out and state on the label that the contents inside the container are dangerous or lethal, you'd better believe it!

Alternatives to Pesticides.

Long before modern chemistry gave us such "marvels" as DDT, gardeners were using natural and organic controls to combat destructive insects and household pests. Organic methods aren't always successful the first time around and often must be repeated to achieve the desired result. But, then, neither are they dangerous to your health or harmful to the environment.

Ants can be trapped by spreading a layer of petroleum jelly (Vaseline) on strips of cardboard where they enter or, often, discouraged from invading your home by spreading tansy leaves around cracks and crevices they are using to gain entry. Stickem works well also.

Aphids can be lured to their destruction by painting the inside of a pie tin or coffee can bright yellow and filling the container with soapy water. Aphids are attracted to the color yellow and will land in the soap solution and be drowned.

A spray/dip concocted of chopped red peppers (hot!), garlic and soapy water will eliminate a colony of aphids in minutes. None of the ingredients harms the plant.

Root aphids can be summarily dealt with either by dumping the contents of a bag of Bull Durham tobacco on the topsoil and watering in, or by soaking the bag in a can containing one cup of boiling water, then pouring the brew over the soil, after it has cooled. Tobacco extracts in this diluted form will not harm plants.

Cockroaches and silverfish can be temporarily eliminated by using one of several nonpoisonous dusts. Two which are very effective and can be used around dishes and foodstuffs are silica gel (a mineral substance) and boric acid (prepared from borax).

Fleas on pets and pet bedding can be controlled without resorting to flea collars, which work by releasing a deadly vapor (DDVP), or potentially deadly sprays. Both pyrethrum and rotenone kill fleas and both are plant derivatives. Some flea powders are made with pyrethrum and rotenone as their active ingredients.

Red spider mites can be eradicated with a spray/dip solution composed of wheat flour, buttermilk and soap. After a few hours, rinse the plant with clear water.

Scales and whiteflies can be killed with rotenone and pyrethrum—scales, before they form their protective shells; and whiteflies, anytime. For best results, repeat the treatment weekly for a month.

Chapter 7

preventive medicine

Children are immunized before they get polio. Adults are given influenza inoculations before they come down with the virus. And dogs are vaccinated before they can be infected with rabies. All of this is preventive medicine. The concept of taking action *before* a crisis strikes is a sound one and should be adapted to your horticultural pursuits.

Give Plants What They Need.

If a plant requires a particular soil mix, light intensity, temperature level, etc., give it what it needs. Don't take shortcuts or develop the arrogant attitude that the plant will see things your way, or else. The "or else" is usually a dire consequence, like the death of a beautiful plant, and this may be the measure your plants take if you try to impose your will on them.

Keep Plants and Pots Tidy.

Potential breeding places for disease and other problems can be eliminated by sprucing up your plants and pots when they need it. Deteriorated leaves should be pruned off and faded or dead blossoms removed. Dead leaves and other debris should be picked off the topsoil. And dust and other obstructions which can block the stomata should be periodically cleaned off foliage with a damp cloth.

Don't Forget the Saucers and Decorative Containers.

Every three or four months, and especially in the case of a sick or infested plant, give saucers and containers a disinfecting bath of Clorox and scalding water. Pests in the pre-

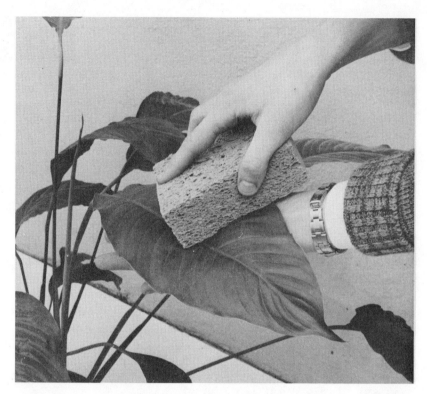

Clear the dust from leaves with a damp cloth or sponge periodically.

emergent stage, and bacteria and fungi can sift down through the soil, out the drainage hole and into saucers and containers where they can reinfest plants.

Isolate Sick or Recuperating Plants.

Plant patients should be separated from your other plants until completely cured. Don't risk spreading the contagion through your entire collection. If practical, moving sick plants to another room is safest. Remember, always wash your hands with soap and water after handling a sick plant. Fungi and pests are easily transmitted to healthy plants, which have no resistance at all.

Don't Prolong the Inevitable.

Develop a "callous" attitude toward plants so badly infested or diseased that, even if they survive, they do only as shadows of their former selves. Accept the fact that you'll have some failures. Keeping terminal cases around as painful reminders is needlessly masochistic. Give your attention to preserving the health of your other plants.

glossary of horticultural terms

Alga (AL'jeh), pl. **algae** (AL'gee) Simple celled plants containing chlorophyll. The fungus form of alga can be seen as the green scum which forms on clay (unglazed) pots.

Axil, pl., **axes** The angle formed where a branch meets the stem of a plant.

Botrytis A group of parasitic fungi which cause plant diseases. Also, a disease caused by a particular fungus.

Callus In a cutting or wound, the new covering of protective tissue which forms to seal the cut and prevent loss of sap.

Canker A swollen protuberance or lesion on a woody stem caused by disease organisms.

Chelating agent (KEE'late·ing) A chemical which makes an element, such as iron, immediately available to a plant's system, so that a plant can get immediate benefit and correct a deficiency before further deterioration takes place.

Chlorophyll The green coloring matter in plants.

Chlorosis An unhealthy yellowing of the foliage of a plant caused by a deficiency of chlorophyll.

Clone A plant which was produced by vegetative propagation of a donor plant. Plants grown from seed, therefore, can never be termed "clones."

Complete fertilizer One that contains the three major nutrients nitrogen, phosphorus and potassium (or potash).

Contact poison A poison that is effective upon contact, as compared with a stomach poison, which must be swallowed.

Crown In discussing a root system, the top and center, from which stems and leaves develop. In plants and trees, the top or head.

Cultivar A botanical variety of a species which originated and is maintained only under cultivation.

Cutting A leaf or stem section of a plant which is taken for propagation of a plant (clone) identical to the parent.

Defoliation In evergreen species, the premature or unnatural falling off of leaves caused by adverse environmental conditions, disease or insect problems.

Degradable A chemical or material which may be reduced in composition or strength. For example, a degradable insecticide is one that eventually loses its "kill potential" or is weakened by time, dilution, decomposition or dispersion in the soil.

Epidermis The outer "skin" or surface of a leaf.

Exuvium, pl., **exuvia** The cast or molted skin of a larva or nymph.

Facultative saprophyte An organism, usually a fungus, that normally exists on dead organic material but can also live as a parasite on living plants.

Fungicide A material that either destroys or inhibits the growth of fungi.

Gall A swelling or growth on a plant or tree caused by an invasion of insects, fungi or other parasitic organisms.

Genus (GEE′nus), pl., **genera** (JEN′er·uh) A plant group, or family, consisting of closely related species having similar characteristics. The first word in a plant's botanical name is the *genus*, the second word indicates the kind or *species* name. For example, *Ficus* is the genus to which all figs belong and *benjamina* tells you what kind of *Ficus*.

Haustorium, pl., **haustoria** The root-like sucker of parasitic plants.

Honeydew A sweet, tacky substance excreted by most sucking insects.

Hyphae The threadlike filaments that make up the vegetative part of a fungus.

Insect vectors Certain insects which carry and spread disease-causing microorganisms. Some, like ants, carry predatory insects to infest healthy plants.

Larva, pl., **larvae** The immature or grub stage of an insect.

Leaching In clay (or terra cotta) pots the action of moisture and minerals passing through the porous sidewalls of the pot.

Loam Any soil that is composed of nearly equal parts of silt and sand and less than a quarter clay. Moistened loam holds together when formed and pressed together.

Microclimate Literally, a smaller climate inside a larger climate, both of which are different. For example, a plant whose pot is sitting on water-covered pebbles is living in a microclimate in which a higher humidity level is produced by moisture evaporating from the pebbles. Terraria and bottle gardens are good examples of microclimates.

Microorganisms Living organisms of microscopic size. Bacteria and protozoa are prime examples.

Midrib The main or central vein of a leaf.

Mildew A fungus that produces a thin whitish covering on the surface of the host plant.

Molt The shedding of skin or other covering in the process of growth.

Mosaic virus A virus which produces a mosaic pattern or mottling of the foliage.

Necrotic In plant tissue, dead brownish areas where cells have been destroyed.

Nematode Microscopic, wormlike animals which cause knots and swellings on roots which distort and stunt plant growth.

Node The joint on a stem or branch where leaves and buds originate.

Nymph The young of certain insects.

Petiole The stem of a leaf.

Pathogene A disease-producing organism.

Photosynthesis The process, still not fully understood, by which a plant's foliage uses sunlight as energy (with chlorophyll as the catalyst) to manufacture sugars and starches from carbon dioxide absorbed from the air, along with water and inorganic salts. The word inself is derived from two Greek words: *Photo*, meaning "light" and *Synthesis*, meaning "to put together."

Pot-bound A condition in which the roots of a plant have become densely packed or have outgrown their container and are unable to function normally.

Potting on Moving a plant to a larger container to provide adequate room for continuing root development.

Respiration The process by which plants assimilate oxygen and expend waste products (comparable to breathing in man and animals).

Ring spots Circular markings on foliage caused by fungi, viruses or other conditions.

Sclerotium The black resting body of certain fungi.

Stem mother A female aphid which gives birth to living young without having been fertilized by a male.

Stoma, pl., **stomata** Minute orifices, or pores, in the epidermis of leaves which enable a plant to carry on respiration and transpiration.

Succulent Any plant capable of storing water in its tissues. Cacti and jade plants are good examples.

Systemic Affecting the entire system. For example, a systemic insecticide is taken up by the roots of a plant and carried throughout the plant's system, from the feeder roots to the tips of the foliage.

Tolerant carriers Plants that carry certain diseases but are not affected by them.

Transpiration A naturally regulated process in which excess water vapor is expended by the leaves of a plant through the stomata.

Vector A carrier of disease-producing organisms.

Vegetative propagation Propagation by using a piece of a plant, rather than seeds or spores.

Wetting agent A substance, usually a soap or detergent, added to the insecticide to improve its spreading and clinging properties, particularly on waxy or glossy leaves.